SpringerBriefs in Education

More information about this series at http://www.springer.com/series/8914

James Deehan

The Science Teaching Efficacy Belief Instruments (STEBI A and B)

A Comprehensive Review of Methods and Findings from 25 Years of Science Education Research

 Springer

James Deehan
Faculty of Arts and Education
Charles Sturt University
Bathurst, NSW
Australia

ISSN 2211-1921   ISSN 2211-193X   (electronic)
SpringerBriefs in Education
ISBN 978-3-319-42464-4   ISBN 978-3-319-42465-1   (eBook)
DOI 10.1007/978-3-319-42465-1

Library of Congress Control Number: 2016945795

Printed on acid-free paper

This Springer imprint is published by Springer Nature
The registered company is Springer International Publishing AG Switzerland

Contents

Acronyms

ES	Effect Size
ICT	Information Communication Technology
KLA	Key Learning Area
NOS	Nature of Science
PBL	Problem-based Learning
PCK	Pedagogical Content Knowledge
PISA	Programme for International Student Assessment
PSTE	Personal Science Teaching Efficacy beliefs
STEBI-A	Science Teaching Efficacy Belief Instrument A (In-Service Teachers)
STEBI-B	Science Teaching Efficacy Belief Instrument B (Pre-Service Teachers)
STEM	Science Technology Engineering Mathematics
STOE	Science Teaching Outcome Expectancies
TIMSS	Trends in International Mathematics and Science Study

Abstract

Science education is currently undergoing a transformation. Students' science interest and achievement levels are waning despite the increasing importance of scientific literacy. Globalisation and technological advancement have raised the bar for economic participation. Therefore, it has become imperative to assess the existing body of science education research. Science education, like all forms of education, is innately difficult to evaluate as achievement outcomes are challenging to quantify and variables remain complicated to account for and control. Amidst a plethora of measures, teacher efficacy has been shown to have positive relationships with teacher resilience, reported use of student-centred teaching strategies and student outcomes. The Science Teaching Efficacy Beliefs Instruments A and B (STEBI-A/STEBI-B) were published 25 years ago as valid and reliable measures of the science teaching efficacy of both pre-service and in-service teachers. Both instruments have become pillars in science education research with a combined citation rate of over 1,400. The value of the STEBI instruments cannot be overstated as both allow for research comparisons across teaching contexts (in-service and pre-service), historical contexts (25 years of research) and national contexts (more than 10 contributing nations). The purpose of this Springer Brief is to provide a comprehensive review of the both the STEBI methods and findings through the use of a clearly defined analytic framework. A systematic review of literature yielded 107 STEBI-A research items and 140 STEBI-B research items. The STEBI instruments have been used in a wide range of qualitative, cross sectional, longitudinal and experimental designs. Analysis of the STEBI research findings reveals that in-service and pre-service programmes which use innovative practices such as cooperative learning, inquiry-based investigation and nature of science instruction can produce positive growth in participants' science teaching efficacy beliefs. The personal science teaching efficacy beliefs of pre-service and in-service teachers showed greater mean scores and higher growth than their outcome expectancies. The implications of the findings in this review are discussed fully in the book.

Chapter 1
Introduction

Much has been said about the crucial role that science has played, and will continue to play, in shaping our societies. For most of us, science conjures images of keystone events in history that have advanced the sum of human knowledge. We consider Charles Darwin's work with the Galapagos Finches, the Moon landing, the discovery of water on Mars and, most recently, evidence indicating the existence of gravitational waves and we marvel at the capacity of human ingenuity. Yet, despite our societies' apparent value of science, it is hard not to feel a sense of disconnect. We didn't form the theories, we didn't conduct the research and we didn't perform the analyses. The unseen barrier between the scientific community and the rest was summed up succinctly by a peer during my undergraduate studies "I can see the value of science but I'm not a <u>sciency</u> person". Upon reflection years later, I came to see this statement as an indictment on this person's fundamental science educational experiences. It was disheartening to hear that a person, with the intellectual capacity to study at the tertiary level, had categorised their scientific disengagement as an innate personal characteristic. This cannot be accepted, science must be for all. The scientific discoveries made by the elite few must be reconciled with the beliefs and understandings of the many; for that is the structure of most truly democratic nations.

Science education is of paramount importance if scientific innovations and discoveries are to continue having meaningful, positive impacts on cultures and societies. In a search for a perspective from an alternate viewpoint, I once asked my mother what she believed the purpose of science education to be. Sensing that I may have been engaging her in an unwanted slow-walk to my own conclusion, she replied with a wry smile, "To train scientists of course!". Her brazen oversimplification had the desired effect as I smiled, sipped my coffee and changed the topic of conversation. The fundamental purpose of science education is to develop scientifically literate citizens. Scientific literacy is comprised of more than the rote recall of existing knowledge; it is a malleable array of skills and beliefs that are generalisable to real-world situations (Bybee 1997). In a practical sense a scientifically literate person would be able to access and interpret information from various sources to reach reasoned, evidence based conclusions whilst avoiding

© The Author(s) 2017
J. Deehan, *The Science Teaching Efficacy Belief Instruments (STEBI A and B)*,
SpringerBriefs in Education, DOI 10.1007/978-3-319-42465-1_1

common pitfalls of logical fallacies. The need for scientific literacy is evident throughout most areas of modern society. From an economic perspective job requirements have changed over the past half a century to more closely resemble the cognitive processes of scientific literacy. Coding and analysis of Census data from the United States over the past 4 decades has shown that Expert Thinking and Complex Communication are the only two skills to have grown in terms of workforce composition (Levy and Murnane 2006). Even environmental issues such as climate change require citizens to both navigate and respond to abundant information and perspectives. It is through the lens of scientific literacy that science education should be judged.

Most citizens will develop their base level of scientific literacy through elementary and secondary science education experiences. Aside from achievement scores, student engagement with science is dwindling. It could be argued that, rather than captivating students, science education can have a filtering effect. This filtering effect begins at the elementary level where students bemoan the passive, note taking pedagogies associated with science classes (Goodrum et al. 2001; Goodrum and Rennie 2007). This is certainly not a new phenomenon as negative science attitudes in students have been reported for decades (Breakwell and Beardsell 1992; Brown 1976; Doherty and Dawe 1988). We are now at a pivotal time as generational cycles of science disengagement could become more deeply rooted in our cultures. According to DeWitt et al. (2014) students' attitudes towards science reaches a steep tipping point at the age of 14 where their engagement with science becomes even more difficult to salvage. This is reflected in reduced rates of post-compulsory science uptake in secondary science education as students elect to separate themselves from science completely (Abraham 2013; Ainley et al. 2008; Hughes et al. 2012; Lyons and Quinn 2010; Osborne et al. 2003). Thus we have the beginnings of the societal disconnect from science that was discussed earlier.

The Trends in International Mathematics and Science Study (TIMSS) and the Program for International Student Assessment (PISA) allow for the science achievement of students to be assessed at global levels. Nations such as America, Australian, England, Japan and New Zealand have all show stagnation and/or decline in the science achievement and scientific literacy of their 4th grade students over the past decade (Martin et al. 1997, 2004, 2008, 2012; Thomson et al. 2008, 2011). TIMSS and PISA data taken from secondary students show similar patterns of decline both in terms of mean scores and international rankings for the aforementioned nations (Beaton et al. 1996; Lemke et al. 2001; Martin et al. 2004, 2008; OECD 2004, 2007, 2010, 2013). Copious numbers of elementary and secondary students appear unable to connect their classroom science learning with the world around them. An undeniable truth has emerged for the nations whose students are performing below high threshold standards; science education is failing in its goal to develop a scientifically literate populace. The collective failure of science education notwithstanding, there is still "hope" as the inquisitive, curious nature of youth cannot be fully extinguished. Indeed many elementary students express a desire to learn more science even though they are seldom active participants in the inquiry process and do not view science as a part of their everyday lives (ACARA 2013).

Finding solutions to the issues with science education is no simple task. The sheer array of stakeholders, variables and potential intervention programs is behemoth. While students represent the key target for change and growth, they remain primarily passive within science education systems. Teachers must play a key role in providing inclusive science education experiences that enhance the scientific literacy of their students. Evaluating preservice and inservice teachers has been an academic and political hot potato for decades and this is not a discussion I intend to contribute to here. One viable construct, among many, is teacher self-efficacy. Bandura (1977) linked an individual's efficacy beliefs with the initiation and sustainment of coping behaviours when experiencing adverse conditions. Gibson and Dembo (1984) developed a valid and reliable instrument for measuring teacher efficacy. Teacher efficacy can be defined as the confidence an individual has in themselves or their profession to help students to achieve pre-determined educational outcomes. The seminal work of Gibson and Dembo (1984) established links between teaching efficacy and other variables such as: use of student centred pedagogies, lower rates of teacher criticism and persistence in difficult professional circumstances. Ghaith and Yaghi (1997) found that teachers with high personal teaching efficacy beliefs were more likely to have open attitudes towards the implementation of new instructional practices. Teacher efficacy has made the intangible tangible as researchers now have a reliable and valid measure for a construct that has established links to many of the factors that contribute to an educational environment. Over the decades efficacy has become a cornerstone for educational research as variable relationships have been established, interventions have been assessed and more efficacy instruments have been developed (Dellinger et al. 2008; Ritter 1999; Tschannen-Moran and Woolfolk Hoy 2001; Smolleck et al. 2006).

For the field of science education, the Science Teaching Efficacy Belief Instruments A and B (STEBI-A/B), for inservice and preservice teachers respectively, have been research linchpins for over 25 years (Enochs and Riggs 1990; Riggs and Enochs 1990). The STEBI instruments have proven to be valid and reliable measures of both personal and general science teaching efficacy beliefs across a variety of contexts (21 contributing nations to be precise!). I was first introduced to the STEBI instruments by my Honours/Doctoral supervisors, Professor David McKinnon and Doctor Lena Danaia, as an overenthusiastic undergraduate research trainee in 2011. David and Lena had been using the STEBI-B instrument in an action research model to evaluate, and subsequently refine, a science program for preservice elementary teachers. I trawled through the STEBI literature and was awestruck by the variation in contexts, purposes, approaches and research designs. A 25-year body of literature seemed endless, but perhaps in a display of the naivety of a neophyte researcher I felt compelled to conquer it. I struggled to reign in my STEBI searching, writing and discussion. There was always more. More viewpoints, more findings, more contradictions, more interventions, more contexts and, above all, more ideas. I soon realised that a thorough review of the STEBI literature needed to be conducted or the messages would soon be lost in the sheer, and increasing, volume of contributions. My apologies for the clichéd, lazy but nonetheless accurate metaphor in advance. I felt that I was holding and using a series of puzzle pieces in isolation without ever having put the puzzle together to see the big picture.

The big picture is now available for you. This Springer Brief presents an unprecedented review of 25-years of STEBI-A and STEBI-B literature (combined citation rate of over 1400). A structural framework is used to analyse and organise the STEBI literature in terms of research designs and findings. This two-phase analytic approach has allowed for the diverse array of STEBI research to be analysed. You will find analyses on the research contexts, subscale use and effect sizes of the STEBI instruments. The STEBI-B research designs are described and critiqued at different levels. Rather than attempting to make broad statements on the quality of the research analysed, the levels have been determined based on the number of STEBI uses within the research design. Such an approaches allows for the unique purposes and contexts to be considered, as comparisons are made where appropriate, rather than universally.

The purpose of this book is not to provide solutions to the issues with science education. That is a task for those with more expertise than I. Over the past quarter of a century many dedicated and insightful researchers have made meaningful contributions to the sustained improvement of science education. Thus, the purpose of this book is to bring together these isolated contributions to form a single, collective overview of how STEBI research has advanced the sum of human knowledge and affected meaningful change across different science education contexts. For too long I, much like many early career researchers, have "stood on the shoulders of giants". *Glances knowingly at the Google Scholar homepage* Perhaps it is time, with the rapid influx of research across emerging platforms, for the giants to be elevated once more.

I urge you, the reader, to extract what you need from this book. For the researchers, the convenience, clarity and power of the reviews (247 articles and dissertations) cater for the demands of PhD students all the way through to established, late career science education researchers. This Springer Brief provides the information needed to make informed research choices. The deep referencing throughout the document can act as a hub for readers to access existing research that suits their needs and interests. For the practitioners (lectures, teachers, subject writers and policy makers), the summary of intervention outcomes can assist you to make informed choices about the design of pre-service and in-service science education programs. The strength of 25 years of comparable literature builds a compelling argument for rich, complex and student-centred science teacher education programs. The definitions of pedagogical innovations, research summaries and deep referencing assist in making this field of research accessible to practitioners and decision makers. This STEBI review could have a tangible impact on how science education programs are designed and delivered. However, a tangible impact is entirely dependent on how you, the reader, respond to the information presented in the following chapters. Let us promote scientific literacy one step at a time.

References

Abraham, J. (2013). Preparing for the future by repairing now: Retaining students in senior secondary physics. *Curriculum and Leadership Journal, 11*(9) (published online).

Ainley, J., Kos, J., & Nicholas, M. (2008). Participation in science, mathematics and technology in Australian education. *ACER Research Monographs, 4.*

Australian Curriculum, Assessment and Reporting Authority (ACARA). (2013). *National assessment program—science literacy year 6 report 2012* (pp. 1–118). Sydney: ACARA. Retrieved from http://www.nap.edu.au/verve/_resources/NAP-SL_2012_Public_Report.pdf

Bandura, A. (1977). Self-efficacy: Toward a unifying theory of behavioural change. *Psychological Review, 84*(2), 191–215.

Beaton, A. E., Martin, M. O., Mullis, I. V., Gonzalez, E. J., Smith, T. A., & Kelly, D. L. (1996). *Science achievement in the middle school years: IEA's third international mathematics and science study.* Boston: TIMSS International Study Centre.

Breakwell, G. M., & Beardsell, S. (1992). Gender, parental and peer influences upon science attitudes and activities. *Public Understanding of Science, 1*(2), 183–197.

Brown, S. (1976). *Attitude goals in secondary school science.* Stirling: University of Stirling.

Bybee, R. (1997). *Achieving scientific literacy: From purposes to practices.* Westport: Heinemann.

Dellinger, A. B., Bobbett, J. J., Olivier, D. F., & Ellett, C. D. (2008). Measuring teachers' self-efficacy beliefs: Development and use of the TEBS-Self. *Teaching and Teacher Education, 24*(3), 751–766.

DeWitt, J., Archer, L., & Osborne, J. (2014). Science-related aspirations across the primary–secondary divide: Evidence from two surveys in England. *International Journal of Science Education*, (ahead-of-print), 1–21.

Doherty, J., & Dawe, J. (1988). The relationship between development maturity and attitude to school science. *Educational Studies, 11*, 93–107.

Enochs, L. G., & Riggs, I. M. (1990). Further development of an elementary science teaching efficacy belief instrument: A preservice elementary scale. *School Science and Mathematics, 90*(8), 694–706.

Ghaith, G., & Yaghi, H. (1997). Relationships among experience, teacher efficacy, and attitudes toward the implementation of instructional innovation. *Teaching and Teacher Education, 13*(4), 451–458.

Gibson, S., & Dembo, M. H. (1984). Teacher efficacy: A construct validation. *Journal of Educational Psychology, 76*(4), 569–582.

Goodrum, D., Hackling, M., & Rennie, L. (2001). *The status and quality of teaching and learning of science in Australian schools.* Canberra: Department of Education, Training and Youth Affairs.

Goodrum, D., & Rennie, L. (2007). *Australian school science education—National action plan 2008–2012* (Vol. 1). Canberra: Commonwealth of Australia.

Hughes, M., Howard, W., Prasad, S., White, S., & Kusa, L. (2012). *Health of Australian science.* Australian Government: Office of the Chief Scientist, Canberra.

Lemke, M., Lippman, L., Bairu, G., Calsyn, C., Kruger, T., Jocelyn, L., & Williams, T. (2001). *Outcomes of learning: Results from the 2000 program for international student assessment of 15-year-olds in reading, mathematics and science literacy.* Washington: National Centre for Education Statistics.

Levy, F., & Murnane, R. J. (2006). Why the changing American economy calls for twenty-first century learning: Answers to educators' questions. *New Directions for Youth Development*, 53–62.

Lyons, T., & Quinn, F. (2010). *Choosing science: Understanding the declines in senior high school science enrolments.* Armidale, NSW: University of New England.

Martin, M. O., Mullis, I. V., Beaton, A. E., Gonzalez, E. J., Smith, T. A., & Kelly, D. L. (1997). *Science achievement in the primary school years: IEA's third international mathematics and science study (TIMSS)*. Boston: TIMSS International Study Centre, Boston College.

Martin, M. O., Mullis, I. V., & Foy, P. (2008). *TIMSS 2007 international science report: Findings IEA's trends in internationals mathematics and science study at the fourth and eighth grades*. Boston: TIMSS & PIRLS International Study Centre.

Martin, M. O., Mullis, I. V., Foy, P., & Stanco, G. M. (2012). *TIMSS 2011 international results in science*. Boston: TIMSS & PIRLS International Study Centre.

Martin, M. O., Mullis, I. V., Gonzalez, E. J., & Chrostowski, S. J. (2004). *TIMSS 2003 international science report: Findings from the IEA's trends in international mathematics and science study at the fourth and eighth grades*. Boston: TIMSS & PIRLS International Study Centre.

OECD. (2004). *Learning for tomorrow's world: First results from PISA 2003*. Washington: National Centre for Educational Statistics.

OECD. (2007). *PISA 2006—Science competencies for tomorrow's world: Volume 1: Analysis*. Washington: National Centre for Education Statistics.

OECD. (2010). *PISA 2009 results: Executive summary*.

OECD. (2013). *PISA 2012 results: What students know and can do—Student performances in mathematics, reading and science* (Vol. 1). OECD Publishing. doi:http://dx.doi.org/10.1787/9789264201118-en

Osborne, J., Simon, S., & Collins, S. (2003). Attitudes towards science: A review of the literature and its implications. *International Journal of Science Education, 25*, 1049–1079.

Riggs, I. M., & Enochs, L. (1990). Toward the development of an elementary teachers' science teaching efficacy belief instrument. *Science Education, 74*, 625–637.

Ritter, J. M. (1999). *The development and validation of the self-efficacy beliefs about equitable science teaching and learning instrument for prospective elementary teachers* (Doctoral dissertation, The Pennsylvania State University).

Smolleck, L. D., Zembal-Saul, C., & Yoder, E. P. (2006). The development and validation of an instrument to measure preservice teachers' self-efficacy in regard to the teaching of science as inquiry. *Journal of Science Teacher Education, 17*(2), 137–163.

Thomson, S., Hillman, K., Wernet, N., Schmid, M., Buckley, S., & Munene, A. (2011). *Monitoring Australian year 4 student achievement internationally: TIMSS and PIRLS 2011*. Melbourne: Australian Council for Educational Research.

Thomson, S., Wernet, N., Underwood, C., & Nicholas, M. (2008). *TIMSS 2007: Taking a closer look at mathematics and science in Australia*. Melbourne: Australian Council for Educational Research.

Tschannen-Moran, M., & Woolfolk Hoy, A. (2001). Teacher efficacy: Capturing an elusive construct. *Teaching and teacher education, 17*(7), 783–805.

Chapter 2
A Review of the Science Teaching Efficacy Belief Instrument B: Pre-service Teachers

Abstract In a world undergoing rapid social, cultural, economic and environmental changes it is imperative to have an informed populace that is capable of displaying the scientific literacy needed to contribute to informed decision making. It is the people, rather than the scientists, that will decide our futures. Tumultuous times call for strong fundamental science education. As the Baby Boomer generation edges towards retirement, it is worth supporting and assessing our future generations of science teachers. Teacher efficacy is a viable means of conducting such assessment as it has shown to relate to pedagogical choices, teacher resilience and student outcomes. The Science Teaching Efficacy Belief Instrument B (STEBI-B) was initially published in 1990 and since this time has proven to be a valid and reliable measure of the science teaching efficacy beliefs of pre-service teachers. The purpose of this chapter is to review the STEBI-B instrument in terms of both methods and findings. Additionally, a framework for the systemic analysis of the literature is presented. A total of 140 articles, dissertations and presentations were included in the analyses. Findings show considerable research design variation. A plethora of student centred science interventions have shown to increase pre-service teachers' science teaching efficacy beliefs. Pre-service teachers' personal science teaching efficacy beliefs consistently show high scores and growth than their outcome expectancy beliefs. Implications are discussed within the chapter.

The Science Teaching Efficacy Belief Instrument—B

The Science Teaching Efficacy Belief Instrument B (STEBI-B) is a 30-item survey which was specifically designed to measure the science teaching efficacy of pre-service elementary school teachers (Enochs and Riggs 1990). This survey requires respondents to rate their level of agreement with statements on a 5 point Likert scale (Burns 2000), ranging from 'strongly disagree' to 'strongly agree'. The statements produce measurements of two subscales. The Science Teaching Outcome

© The Author(s) 2017
J. Deehan, *The Science Teaching Efficacy Belief Instruments (STEBI A and B)*,
SpringerBriefs in Education, DOI 10.1007/978-3-319-42465-1_2

Expectancy (STOE) belief scale measures the participants' broad views of science teaching related to why pupils perform as they do. An example of an item on the STOE subscale is "when a student does better than usual in science, it is often because the teacher exerted a little extra effort". The Personal Science Teaching Efficacy (PSTE) scale measures the participants' beliefs about their own ability to teach science effectively. An example of an item on the PSTE subscale is "even if I try very hard, I will not teach science as well as I will most subjects".

Although other instruments, such as the Self-Efficacy Beliefs About Equitable Science Teaching (SEBEST) instrument (Ritter 1999), have been developed, the STEBI-B is frequently used within the science education research domain due to its capacity to measure relevant, complex constructs in a reliable way. When Enochs and Riggs created the STEBI-B instrument in 1990, the PSTE and STOE subscales were found to have Cronbach Alpha reliability coefficients of 0.90 and 0.76 respectively. A recent investigation (Deehan 2013) found that the STOE produced a Cronbach's alpha of 0.798 which appears to, in a small way; quell the growing doubts about the reliability of the STOE subscale (Hechter 2010; Johnston 2003; Mulholland et al. 2004; Ramey-Gassert et al. 1996; Watters and Ginns 2000; Yilmaz and Cavas 2008). The reliability of the STEBI-B will be unpacked further in the discussion section of this chapter.

The Research Contexts of the STEBI-B Instrument

After a relatively limited research uptake in the 1990s (16 studies) the use of the STEBI-B instrument increased by approximately 900 % in the new millennium. This could be partially attributed to: increased awareness of the instrument internationally, an increased interest in exploring the problems associated with science education and the growth in options for publishing research globally.

The majority of the STEBI-B research originates from the USA. Significant amounts of research have also been emerging from Australia and Turkey. The differences in social, cultural and educational factors amongst these nations provide worthwhile checks and counterbalances for the American research findings. In fact, researchers themselves have recognised potential in such collaboration, as various formal connections between the three key nations are present within the body of literature (e.g. Çakiroglu et al. 2005; Rogers and Watters 2002). These three nations (USA, Turkey and Australia) account for 91.4 % of the current STEBI-B research. Unfortunately, there does not appear to be any other nation on the cusp of matching the research contributions of the main nations. There may be opportunities for more international collaboration as researchers from Austria, the Bahamas and Greece have made meaningful contributions to the STEBI-B literature since 2014. Figure 1.1 compares the STEBI-B research output internationally. Aside from following a similar pattern of increased research output after 2000, the global research is appearing sporadically.

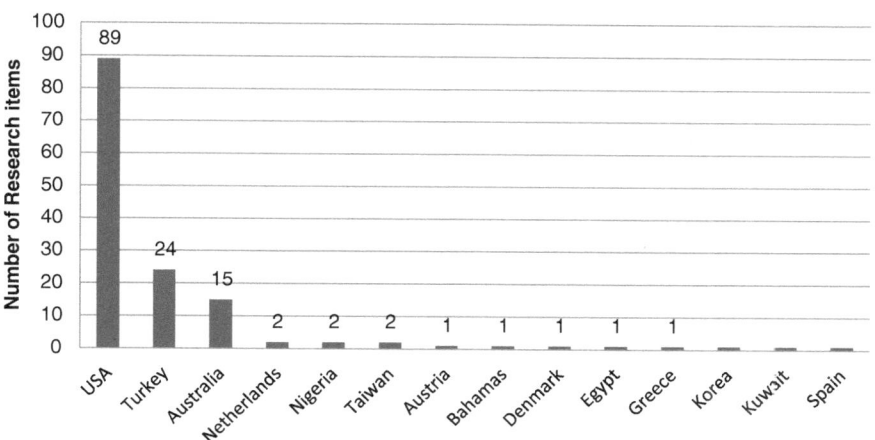

Fig. 1.1 Summary of the STEBI-B use of different nations

Purpose of This Study

There are four aims that underpin this chapter, with the first aim setting the structure for those to follow. The first aim is to articulate a coherent framework for organising and discussing the STEBI-B literature in a way that considers the inherent complexity of science education research without attaching undue value judgements to researchers' choices. This framework is applied to the STEBI-A review in the second chapter. The second aim is to provide a clear overview of how the STEBI-B is being employed methodologically within the growing body of literature. In a logical progression from context to instrument, the third aim is to describe how the STEBI-B subscales (PSTE and STOE) are being employed within the literature base. The final aim is to compare the effects of different pedagogical approaches undertaken in pre-service teaching programs as outlined via the STEBI-B instrument.

Method

This research has been conducted through the use of a structural framework to comprehensively review the body of STEBI-B literature. The intention of this research is to explore the trends which have emerged within the science education literature that has employed the STEBI-B instrument. Due to the open-ended, inductive nature of review paper, the author has chosen not to list specific research questions as this could limit potential findings and inductive trends. It should be noted that the reliability (Cronbach's alpha) of the STEBI-B instrument itself is not the focus of this paper. The aspects that will be the focus of classification and analytic procedures in this review are: context, research design, interventions, participant numbers, subscale use and Cohen's D effect sizes.

The following subsections will outline the procedures for both how research papers were collected for inclusion in this review and what analytic processes were undertaken to evaluate the literature. The first subsection will outline the inclusion criteria and the literature search techniques. The coding and analysis subsections will present deep explanations of the coding and analytic procedures for the methodological and intervention analyses.

Initial Inclusion Criteria

To adhere to the purposes of this review paper a single broad (Suter 2006) criterion was employed during the initial literature searches. This criterion was that the STEBI-B instrument is used to inform the research in a meaningful way. This accounts for the diverse contexts within which the STEBI-B instrument can be used.

Initial Literature Collection Procedures

The initial search for literature occurred in 4 phases.

1. **Seminal author search**—The article describing the development of the STEBI-B instrument by the seminal authors (Kervin et al. 2006) was found using 'Google Scholar'.

   Enochs, L. G., & Riggs, I. M. (1990). Further development of an elementary science teaching efficacy belief instrument: A preservice elementary scale. *School Science and Mathematics*, *90*(8), 694–706.

   A backward mapping search (Green et al. 2006) was used to track articles that had referenced this seminal article. The seminal author search yielded 255 articles, 111 of those were deemed relevant for this review.

2. **ERIC Search**—The Education Resource Information Centre database was specifically searched with the use of the following terms:

   'STEBI-B'—Two papers of the 20 presented fulfilled the inclusion criteria.

   'Science Teaching Efficacy Belief Instrument B'—Four papers of the 50 presented fulfilled the inclusion criteria.

3. **Primo Search**—The Primo search website was used to search for relevant literature. In addition to the previously mentioned data bases, Primo provides access to journal articles, newspaper articles, books, Ebooks and other forms of research output. The following search terms were employed:

'STEBI-B'—This search yielded no relevant results.

'Science Teaching Efficacy Belief Instrument B'—This search term yielded two additional papers.

4. **Branching off Bibliographies**—The final step of the literature search was to read the introductions, discussions and reference lists (Green et al. 2006) of the collected papers in order to identify articles of research that had not been collected within the previous steps. The most recent papers were searched initially. This strategy yielded no new, relevant STEBI-B articles.

A total of 117 relevant research items were collected during the initial literature collection phase.

Additional Data Base Searches

After consultation with a literature search expert, additional data base searches were employed with the intent to supplement to earlier procedures. The key search terms that were used for each database were 'STEBI-B' and 'Science Teaching Efficacy Belief Instrument B'. The following databases were searched in this phase of the literature search: EBSCO Host, Cambridge Journals, CBCA Database, Emerald, Expanded Academic ASAP, Infotrac, Factiva, Informit, Web of Knowledge, JSTOR, Oxford Journals, ProQuest, SAGE Journals Online, ScienceDirect (Elsevier SD), Scopus, Springerlink, Taylor & Francis Online, and Wiley Online Library. Another 34 relevant pieces of research were acquired through these additional data base searches. At the conclusion of the literature search phase, 151 articles were collected that appeared to meet initial inclusion criteria.

A Complementary STEBI-A Search

The STEBI-B instrument is the pre-service equivalent to the Science Teaching Efficacy Belief Instrument A (STEBI-A) (Riggs and Enochs 1990). Both instruments are based on the same core items, measuring the same subscales, with slight phrasing modifications to suit the separate contexts. As both the STEBI-A and STEBI-B instruments were published by the same authors in the same year, the researcher believed that some relevant STEBI-B research may be misrepresented in the STEBI-A body of literature. The aforementioned literature search strategies were employed for the STEBI-A instrument. A total of 12 STEBI-B research items were found in this manner. This took the total number of STEBI-B articles to 163. The final number was reduced to 140 pieces of research that incorporated the STEBI-B instrument after 23 repeats, alternates and inappropriate articles were removed.

Coding and Analysis

The research items selected for this review were coded in different ways to allow for holistic analyses to be conducted. The coding was used to differentiate the research approaches and the interventions employed. The following subsections will articulate the coding procedures for the research approaches, science interventions and the use of Cohen's D effect sizes to evaluate the science interventions.

Research Approaches

The coding of the research methods employed was centred on the use of the STEBI-B instrument. As a result the frequently discussed balance between qualitative and quantitative approaches was not an overt focus, but emerged sporadically throughout the analyses. The research items were coded based on the number of administrations of STEBI-B instrument, the contexts in which the administrations occurred and how the data was analysed and presented. Table 1.1 below outlines the code descriptions for the different research approaches.

To gain a deeper understanding of the body of literature, the research pieces were coded in different ways where appropriate. Firstly, the subscale differentiation (PSTE and STOE) was coded on a 3-point scale. A score of zero indicated that the subscale was not present, a score of 1 indicated that the subscale was merged, and a score of 2 indicated that the subscale was present. Secondly, the descriptive statistics of the qualifying studies were coded in terms of overall quality, within an underlying focus on the calculation of Cohen's D effect sizes for intra-study comparisons. A score of 2 indicated all necessary statistics have been presented. A score of 1 indicated that some necessary statistics have been presented. A score of 0 indicated that the descriptions were not clearly presented.

Intervention Coding

The interventions employed within the research papers were coded based on the pedagogies included within a science intervention. A framework of pedagogical elements was developed primarily from Lawrance and Palmers' (2003) description of innovative practices within tertiary science programs in conjunction with wider literature reading. The researcher acknowledges that this is not an exhaustive list of potential pedagogical approaches. The argument could certainly be made that several of these innovative practices overlap. This is unavoidable as many are designed to provide students with control over their learning. For example, one could make the claim that constructivism is an integral component of many other innovative practices. There are, however, subtle differences between the innovative practices in terms of pedagogies, contexts, intended learning outcomes and overall

Table 1.1 Research methods codes

Code	Research type	Number of STEBI-B uses	Description
4	Experimental design	>2	Experimental research designs allow for cause and effect statements to be made rather than correlational observations. Two groups comprise this research design. The experimental group is exposed to a formal treatment. A control is does not receive the same treatment. Where possible extraneous variables are controlled through randomised group assignment. However, in educational research it is often ethically impossible to randomly assign participants to either group
3	Longitudinal quasi experimental pre/post design	>2	Research where a pre and post-test STEBI-B implementation is supplemented by delayed testing to determine the longevity of any efficacious changes in the absence of formal science treatment
2.5	Quasi Experimental pre/post—with multiple cohorts	>2	Research where multiple cohorts of participants responded to pre- and post-test versions of the STEBI-B, as they undertook a specified intervention
2	Quasi Experimental pre/post	2	Research where a single cohort of participants provided STEBI-B data, both before and after a specified period of time and/or undertaking a science intervention
1.5	Equivalent groups	2	Research where pre- and post-intervention STEBI-B data were collected from separate groups and compared as equivalent data (Suter 2006)

(continued)

Table 1.1 (continued)

Code	Research type	Number of STEBI-B uses	Description
1	Cross sectional	1	Research where the STEBI-B was administered to a group of pre-service teachers on one occasion to make comparisons with other variables
0	Qualitative/alternate research purpose	Variable	There are two primary types of research that fall into this category; Research where the STEBI-B instrument was not used to provide statistical data, as originally intended by the seminal authors (Enochs and Riggs 1990). Examples include the use of STEBI-B items as the basis for interview questions (e.g. Tosun 2000) and the use of STEBI-B as an instrument for professional reflection (e.g. Lewthwaite et al. 2012). Research where the STEBI-B instrument was outlined in the methodology, but the subsequent STEBI-B data was not presented (e.g. Watters 2007)

focus. Table 1.2 below explains the selected innovative practices. The list of innovative practices provided as a part of the framework is by no means infallible. If anything, this list needs to be refined and modified in the future as science education research continues to progress.

The interventions were coded dichotomously as either including (1) or not including (0) each innovative practice. The judgement was based upon the author's thorough reading of the intervention descriptions, which were supplemented by the use of a search function to assess the use of key terms. An innovative practice did not have to be explicitly explained within an intervention to be classified as 'included', rather the practice had to be evident within the description based on the informed reading of the researcher. The quality and depth of innovative practices were not differentiated in this coding scheme.

Table 1.2 Overview of innovative practices used within the analyses

Innovative practice	Description
Constructivism	Learning that occurs when an individual constructs their knowledge through active participation (i.e. discussion) within a phenomenon or situation (Slavin 1991; Vygotsky 1977)
Problem-based learning	Problem-based learning is a deep learning strategy that helps students to develop transferrable skills, which can be used in novel situations (Schmidt et al. 2006). Problem-based learning uses real-world problems as a starting point for the acquisition and integration of new knowledge into existing schemas (Azer 2001; Kahn and O'Rourke 2005)
Integration with other key learning areas (KLAs)	An approach to teaching where two disciplines, that are considered fundamentally separate, are integrated to create deep learning outcomes. For example, allowing students to collect and graph data is an example of a deep integration between mathematics and science
Mentoring	Mentoring is an emerging practice where pre-service teachers are paired with experienced teachers in order to focus on a particular discipline (e.g. Kenny 2010). The pre-service teachers observe experienced teachers and receive feedback on their own emerging teaching practice
Curriculum development	This term broadly encompasses teaching pedagogies and learning opportunities that accurately reflect the responsibilities and actions of the profession for which the students are being trained to enter. Within the context of this review this would include approaches such as allowing the pre-service teachers to create science units of work for classroom use
Inquiry learning	Inquiry learning allows participants to develop transferable skills and knowledge to seek the information needed in order to achieve a task (Duran et al. 2009; Edelson et al. 1999). Open inquiry occurs when the participants have complete control over processes of inquiry. Guided inquiry occurs when some structure is provided to guide students towards a learning goal
In-subject practical experience	This occurs when the intervention is designed with imbedded opportunities to teach science to students of the intended year levels
Links to practical experience blocks	Unlike 'in subject practical experience' this occurs when the student teachers are required to undertake some form of science teaching and reflect upon their experiences after they complete a specified tertiary science subject

(continued)

Table 1.2 (continued)

Innovative practice	Description
Cooperative learning	Cooperative Learning occurs when students work together in separate, complementary roles to complete a task that would otherwise be impossible to complete individually
ICT instruction/incorporation	The students explicitly learn about the use of ICT in a way that is relevant to classroom teaching practice. For example, the use of Interactive Whiteboard Software for creating learning aids. Deep ICT instruction may be imbedded within subject assessment
Student centred investigation	These are investigations where the students assume the locus of control within the confines of the subject. Ideally, students should have control over all stages of the investigation, with the instructor acting in a facilitative role
Authentic tasks	In the tertiary science context, authentic tasks are those that are clearly related to the profession/career that the students are studying to enter. Examples may include developing units of work, practical experiences and researching student misconceptions
Nature of Science	The understanding that scientific knowledge is fluid and always subject to reasonable debate. Instruction in this area may orient the learner to the variety of scientific approaches beyond an experimental research design
Misconceptions	A misconceptions based approach is a practice where the misconceptions of the students are identified and revealed to them. These misconceptions form the basis of the learning experiences delivered to the students

Using Effect Sizes to Evaluate the Interventions

The effect sizes of STEBI-B studies that included a tertiary science intervention targeting pre service teachers and used the STEBI-B at least twice in a pre/post-test design (i.e. a code 1.5 or greater on the research approaches) were collected and compared to determine which approaches correlated with the strong increases in the PSTE and STOE scores of the participants.

The calculated Cohen's D Effect sizes were selected for research that utilised a single group of participants. However, for relevant research with multiple cohorts an average effect size was calculated to most accurately reflect the science teaching efficacy changes within the participants. For research with two cohorts, a mean was calculated. Within research that included three or more experimental groups, a statistical outlier was classified as an effect size of 0.5 higher or lower than the closest ordinal effect size. When no outliers existed a mean was calculated. When an outlier was present, the median effect size (or mean of the median scores) was selected to represent the research. The final number of research papers and dissertations included in this review was 140.

Organisational Framework for the Analysis of the STEBI-B Literature

My initial conceptualisation of this STEBI-B review was to construct a funnel that systematically critiqued and eliminated research at a variety of levels until the reader was left with most exemplary research that utilised the instrument. In retrospect, such a conceptualisation is a vast oversimplification of a 25-year old body of literature that features contributions from 14 different nations. Simply put, no single funnel exists that can identify 'the best' research. The unique social, cultural and historical contexts that the research articles and dissertations reflect make definitive judgements and comparisons highly complex. I do not have the inclination or the expertise to attach value judgements to the 140 STEBI-B research items from a research design perspective. Instead, I chose to include comparisons in this paper in an objective and consistent fashion. Methodological groups, descriptive statistics, subscale analyses and Cohen's D effect sizes are presented to meet the aims of this paper. As an outsider, I am not privy to the complex interactions that have occurred amongst the reported innovative practices and the subsequent STEB changes of participants. Thus, I now conceptualise my role as organising and critiquing the research without making definitive statements on methodological quality and providing a point of objective comparison of 'innovative' science interventions. It is hoped that the structure presented allows the reader to make his or her own judgements based on his or her unique perspectives.

The organisation and presentation of 140 pieces of individual research demands a definitive structure. Content and purpose cannot be divorced from methodology in a single piece of research. This is the primary reason why this Springer Brief reviews research designs and assesses science interventions. A separation of these elements would leave the story incomplete. 'Research designs' has been selected as the core principle within the organisational framework. The methodological choices of the researchers provide a clear and universal lens through which research can be grouped and discussed. The focus on methodology also helps to enhance the narrative of the paper as the research presented builds towards the evaluative focus of the latter half of the paper. An objective comparison is achieved as dichotomous coding of innovative practices is reconciled with STEB effect sizes. Research designs and innovative practice summaries are provided to discuss emergent trends. No conclusions can be drawn in either domain.

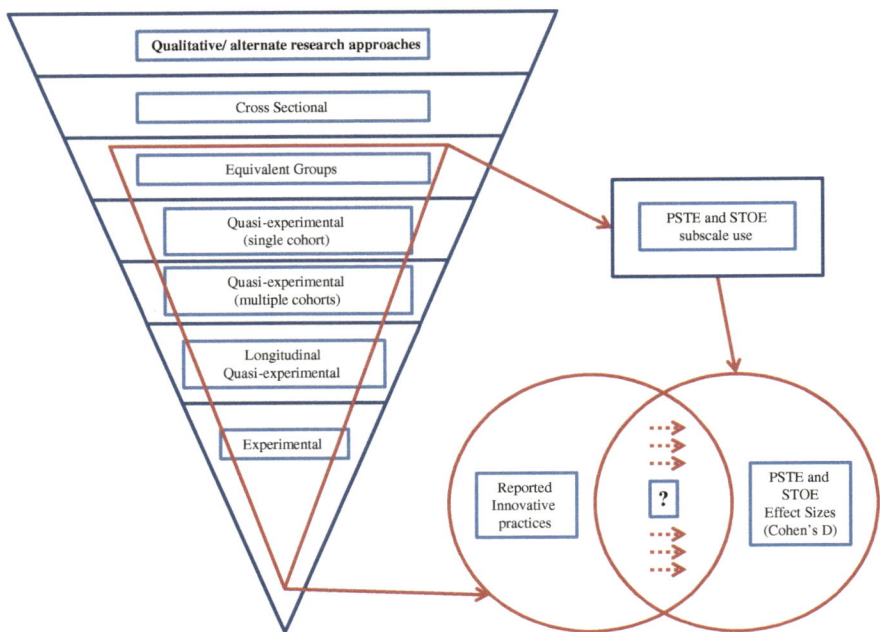

Fig. 1.2 STEBI-B review framework

Figure 1.2 shows the STEBI-B review framework. Blue has been used for the research designs review and Red has been used for the more complex innovations comparison. The blue inverted triangle to the right shows how all research articles have been grouped and separated based on the number of STEBI-B administrations. The research designs at the top of the triangle generally used the STEBI-B on fewer occasions than those at the bottom of the triangle. Once categorised and discussed an individual paper is removed from future sections research design analysis. This is represented by the funnel shape. The red triangle represents a separate analysis of research items that use the STEBI-B at least twice (pre/post test) to evaluate a science intervention. The two red arrows connecting the red triangle to the Venn diagram shows the information that was extracted from the relevant articles to gain a broader understanding of the effects of the different "innovative" practices used within science interventions. The complexity of the research domain is shown by the red arrow that passes through "PSTE and STOE subscale use" before the "PSTE and STOE Effect sizes (Cohen's D)". Research papers needed to employ the recognised subscales using accepted statistical approaches for comparisons to be made. The red circle on the left of the Venn diagram represents the reported innovative practices. The red circle on the right of the Venn diagram represents the reported PSTE and STOE Effect sizes. The intersection of the Venn diagram represents the connection between reported innovative practices and STEB Effect sizes. The dotted arrows and the question mark show the hypothesized, but unknowable, relationship between these variables. Indeed this represents a limitation of the research, as the researcher cannot know the contextual interactions that may have influenced the relationship between these variables.

Findings

The findings for this review are presented in two sections. Firstly, an overview of the use of the STEBI-B instrument within research designs will be presented. This section will be broken down into research designs and subscale usage. Secondly, a comparison of the effects of the different science interventions reported in the STEBI-B literature will be made. This will be achieved by outlining the use of the different innovative practices and then comparing the same papers in terms of PSTE and STOE effect sizes.

STEBI-B Research Designs

The following subsections will present and discuss trends amongst the different research designs employed within the STEBI-B body of literature. At the end of each section the discussed papers will be removed from the analysis. This adheres to the conceptual framework outlined above as the funnel focuses on STEBI-B administrations to lead into the evaluation of innovative practices further along in the paper. After repeats were removed, a total of 140 articles were included in this section of the analysis.

Qualitative and Alternate Research Approaches (0)

Much of the research coded at this level was limited by small sample sizes (Englehart 2008; Lewthwaite et al. 2012; Peters-Burton and Hiller 2013; Soprano and Yang 2013). Englehart's (2008) deep case study approach suggests that inquiry based curriculum support materials can improve the science teaching efficacy, pedagogical content knowledge and inquiry teaching practices of pre-service early childhood teachers. However, due to small number of participants (3) the results cannot be generalised to alternate contexts and the statistics cannot be calculated for comparison. Similarly, the research of Peters-Burton and Hiller (2013) was marred by a low number of participants. The six participants responded to the STEBI-B instrument as a complement to interview data. The findings showed that the pre-service teachers placed a stronger emphasis on how 'fun' a science lesson could be rather than the science concepts being delivered, even when their students directly requested the latter. Soprano and Yang (2013) opted to use the

STEBI-B instrument within a deep, action research paradigm. A single pre-service teacher responded to 13 items on the PSTE subscale and each item was analysed independently.

A number of research projects featured the STEBI-B in a qualitative context (Lewthwaite et al. 2012; Tosun 2000; Watters 2007), thus limiting the relevance in this review. Tosun (2000) used the STEBI-B items to design a series of semi-structured interview questions and found that 44 pre-service elementary teachers had overwhelmingly negative science teaching efficacy. In an attempt to assess the effect of Nature of Science (NOS) instruction, Lewthwaite et al. (2012) used the STEBI-B to aid the personal science teaching reflection of a single pre-service teacher. The presentation of STEBI-B data was another issue that arose in this level of coding. Watters (2007) outlines a strong longitudinal, experimental research design employing a mixed methods approach. However, the STEBI-B data were not presented clearly or consistently within the results. The data could not be classified as transparent as no standard descriptive statistics were provided to allow for cross checking. Although Watters (2007) reported that the intervention produced positive gains within the 360 participants, the aforementioned issues with the results have prevented this study from being included in further analyses.

A recent trend in the STEBI-B literature has been the modification of the STEBI-B instrument for more specific, alternative contexts. Such modifications of the original STEBI-B instrument fundamentally change the targeted constructs, thus the studies have been coded as 'alternate approaches'. Wilson (2012) modified the STEBI-B instrument to focus on pre-service teachers' conceptualisations of sustainability. The participants reported high self efficacy in their capacity to deliver education for sustainability with many believing that they would openly encourage inquiry questions for which they did not have an immediate answer. In a similar focus area, Richardson et al. (2014) chose to adapt the STEBI-B instrument to explore environmental education self efficacy. The new instrument was used across both a science content subject and a science methods subject. The participants' personal efficacy increased in the content course but decreased in the methods course. The outcome expectancies showed no change and were subsequently dismissed. Other STEBI-B modifications were based around: technology efficacy (Ting and Albion 2014); Astronomy concepts (Ivey et al. 2015); Inquiry science teaching (Avery and Meyer 2012); and the need to adapt to new cultural contexts (Park 1996). These alternate research pathways may be a signpost for the future of the STEBI-B body of literature. A total of 24 research papers were either classified as qualitative or deemed to be employing an alternate research approach. After these studies were eliminated, 116 studies were included in the next step of the analysis.

Cross Sectional Research Designs (1)

Research with a cross-sectional usage of the STEBI-B instrument has allowed the construct of science teaching efficacy to be linked to a broad array of variables.

Variables such as classroom management beliefs (Gencer and Çakiroglu 2007), epistemological views (Sunger 2007; Yilmaz-Tuzun and Topcu 2008) and science content knowledge (Mashnad 2008; Sarikaya et al. 2005) have all been analysed in relation to STEBI-B data. More specifically, Gencer and Çakiroglu (2007) utilised a robust sample of 584 preservice science teachers to identify a negative correlation between PSTE and STOE scores and the use of teacher-centred, interventionist classroom management strategies. Sunger (2007) found that both elementary teachers and secondary science teachers expressed moderately high STEBs and viewed the acquisition of knowledge to be underpinned by non-linear reasoning, repeated learning and continued inquiry. Yilmaz-Tuzun and Topcu (2008) explored epistemological views in relation to preservice teacher STEBs in greater depth. Multiple regression analysis showed that preservice teachers with higher STEB scores were less likely to believe that their students' capacity for learning is a fixed, unchanging characteristic. Curiously, some evidence was presented that showed higher STOE scores were related to beliefs that science is composed of fixed, unchanged knowledge. Sarikaya et al. (2005) employed Multiple Regression Correlation Analyses to determine the extent to which STEBs accounted for the variance in 750 Turkish preservice teachers' science knowledge scores. Results showed that PSTE accounted for 40 % of the variance, whereas STOE accounted for just 4 % of the variance. Contrariwise, Mashnad (2008) found no link between the science content knowledge of 91 preservice teachers and their STEBs. One interpretation was that the participants displayed a limited awareness of the alternative science conceptions they continued to hold as adult learners.

Comparisons between mathematics and science views are prominent within the STEBI-B literature (Bursal and Paznokas 2006; Bursal 2010; Wenner 2001). Bursal (2010) found that despite a strong positive correlation between personal mathematics teaching efficacy and personal science teaching efficacy, the respondents had much higher mathematics teaching efficacy scores. Earlier research (Bursal and Paznokas 2006) further strengthens the connection between science and mathematics attitudes amongst pre-service teachers. The data indicated that there was a negative correlation between reported science teaching efficacy and mathematics anxiety.

Cross sectional STEBI-B administration has been used to analyse science teaching efficacy in relation to scientific misconceptions (Schoon and Boone 1998; Tekkaya et al. 2004). A sample of over 600 pre-service elementary teachers revealed that low science teaching efficacy beliefs covaried with reported fundamental science misconceptions (Schoon and Boone 1998). The key misconceptions reported were based on planets, dinosaurs and electricity. Evidently, fundamental gaps in science content knowledge serve as barriers to the development of science teaching efficacy in pre-service teachers. Yet, Tekkaya et al. (2004) found that this covariant relationship does not extend to the secondary teaching domain. A group of 299 pre-service science teachers reported confidence in science teaching despite holding misconceptions concerning fundamental science concepts.

A significant portion of the body of literature employed the STEBI-B to compare the science teaching efficacy beliefs of different sub-groups of pre-service teachers (Arigbabu and Oludipe 2010; Çakiroglu et al. 2005; Newsome 2003; Rogers and Watters 2002; Wenner 2001). Wenner (2001) analysed response rates to the different items on the STEBI-B scale to compare the science teaching confidence and accountability perceptions of pre-service and in-service teachers. Perhaps unsurprisingly, the in-service teachers reported higher science teaching confidence. Nevertheless, pre-service teachers appeared more receptive to student questioning in science. In terms of accountability, only 53 % of pre-service teachers believed that teaching was responsible for student achievement. Newsome (2003) compared the STEBs of various sub-groups of pre-service teachers. In this study, pre-service teachers who completed in-school professional development held higher PSTE scores than their counterparts in traditional tertiary courses. In line with familiar themes in the literature, student teaching setting, academic level, academic major or area of science concentration did not affect the STOE scores.

Many of the STEBI-B research items emerging from Turkey employ the STEBI-B in cross sectional ways (Bahcivan and Kapucu 2014; Kahraman et al. 2014; Olgan et al. 2014; Serin and Bayraktar 2014). Bahcivan and Kapucu (2014) assessed the PSTE scores of 379 pre-service teachers in relation to their conceptions of science learning. The results indicated that investigative constructs were positive indicators of participants' PSTE scores. Curiously, 'application of skills' was a negative predictor of PSTE, suggesting that the participants were not yet comfortable shifting from theoretical understandings to practical science engagement as teachers. Olgan et al. (2014) delved into the issues with the STOE subscale by determining how the construct is influenced by other variables. The results showed that PSTE and justification of science knowledge were sound predictors of STOE scores. Curiously, epistemological beliefs, attitudes toward science teaching and scientific content knowledge did not have significant influences upon the STOE scores of the 379 pre-service teachers. Serin and Bayraktar (2014) researched the relationship between pre-service teachers' beliefs about locus of control and their science teaching efficacy beliefs. The results showed that participants with internal locus of control beliefs had higher science teaching efficacy than their counterparts with external locus of control beliefs. A total of 32 research items, cited within this review, adopted the STEBI-B in a cross sectional way. After these studies were eliminated, the pool of research papers decreased to 84.

Research with Equivalent Groups (1.5)

The research presented within this section of the analysis is similar to the 'cross sectional' research in that each participant is only exposed to the STEBI-B instrument on one occasion. However, the following authors have attempted to circumvent the lack of comparative opportunities within a cross sectional approach by comparing the STEBs of separate, but equivalent, groups that differ on one or more key variables. The average number of participants in research using equivalent groupings is 241 (Aydin and Boz 2010; Bayraktar 2011; King and Wiseman 2001;

Luera and Otto 2005; Velthuis et al. 2014; Wenner 1995). This is a mean of over 100 participants higher than the mean of 125 participants shown in the 129 articles that provided clear information on participant numbers. The mean number of participants for single cohort pre/post designs is 67. It appears as though the researchers using the STEBI-B in an equivalent groups design may be partially overcoming the lack of pre/post case matching with significantly higher numbers of participants.

Equivalent group designs are often used to research science education programs (Wenner 1995; Velthuis et al. 2014). Wenner (1995) used an "equivalent groups" design to assess the effectiveness of changes to a science program implemented over two years. The first sets of data were taken from pre-service teachers in 1992 prior to an increase in the number of science subjects. The follow-up data set was taken from a separate cohort in 1994, after the core changes had occurred. Results indicated that the second group, who experienced the changes, reported significantly higher PSTE scores than the 1992 group. Similarly, Velthuis and others (2014) compared the STEBs of multiple pre-service teacher cohorts between two universities in order to determine the effect of increased mandatory science subjects. A curious finding was that first year pre-service teachers who experienced a science content course reported higher PSTE scores than those who partook in a science methods course. However, by the second year of study this difference had disappeared.

Equivalent groupings can be used to evaluate preparatory teacher education courses rather than single science subjects (Aydin and Boz 2010; Bayraktar 2011; Luera and Otto 2005). Luera and Otto (2005) focused on the impact of the number of science subjects offered on the science-teaching efficacy of pre-service teachers. A group of 20 pre-service teachers who completed the institution's three science subjects were compared to a baseline group of 101 pre-service teachers who had not begun their science studies. Experience with three science subjects covaried with higher PSTE scores (Cohen's D = 0.857). Curiously, there was no significant difference between the groups on the STOE subscale. A similar design was used in a Turkish context to assess the holistic effect of an undergraduate degree on the STEBs of pre-service teachers (Bayraktar 2011). A comparison between the PSTE scores of first and fourth year students suggested that those who had undertaken the course curriculum experienced moderate gains in their personal science teaching efficacy beliefs. Additional research has compared first and fourth year pre-service teachers for a broader programmatic focus (Aydin and Boz 2010). Results showed that the fourth years had higher PSTE scores but very similar STOE scores to their first year counterparts. After the six 'equivalent groups' articles were removed from the analysis, 78 remained for the following section.

Quasi Experimental Designs with a Single Cohort (2)

The quasi-experimental research items almost solely utilised the STEBI-B instrument to explore covariant relationships between participation in different science interventions and STEB changes from pre- to post-occasions of testing. To prevent duplication, the findings of papers at this level will be explored in greater detail later in this chapter. Pre- and post-test administrations of the STEBI-B were used to assess science interventions that included an array of pedagogies including: constructivism (Bleicher and Lindgren 2005); field experiences (Plourde 2002; Sindel 2010; Wagler 2011); inquiry learning (Shroyer et al. 1996); problem-based learning (Wingfield and Ramsey 1999) and misconception targeting (Jabot 2002). Templeton (2007) reported on the science teaching efficacy development of a group of pre-service teachers as they employed constructivist approaches to design science curriculum for a local museum. Setting aside the small sample of 14, the participants displayed a 2-sigma effect size growth in their reported PSTE scores and close to 1-sigma growth on the STOE subscale. A science methods course that afforded participants the chance to teach science, record their science lessons and reflect on their science teaching practice, lead to similar effect size growth in participants' STOE scores (Naidoo 2013). However, such improvements to pre-service teachers' beliefs about the capacity of science teaching to improve student-learning outcomes appear to be outliers within the STEBI-B literature base.

Much of the research at this level of the analysis reports stagnation, and in some instances slight declines in the science-teaching outcome expectancies of pre-service teachers (Bursal 2008; Hudson 2004; Plourde 2002; Watters and Ginns 1999; Yilmaz and Cavas 2008). Plourde (2002) described a science methods course that used constructivist approaches to prepare participants for an imbedded practical science teaching experience. The 59 pre-service teachers showed stagnated PSTE scores and moderate effect size declines in their STOE scores. The author attributed these declines to contextual in-school factors, such as; insufficient time, limited resources and the absence of collegial support, which are commonly cited within the literature (Goodrum et al. 2001; Goodrum and Rennie 2007; Griffith and Scharmann 2008). The stagnation of the PSTE scores was ascribed to participants' negative experiences as students themselves. Later research suggests that Plourde's interpretations may be accurate (Yilmaz and Cavas 2008). The STEBs of 185 pre-service were unaffected by in-school teaching placements, which may be another piece of evidence of the aforementioned issues with science education.

The depth and quality of research using the single cohort, pre-post test design has continued to improve in recent years. Since 2014, 10 studies have been published to make meaningful contributions to the existing STEBI-B literature. The pedagogical base is expanding beyond the limitations of the conceptual framework in this paper, making it difficult to summarise the innovative practices in succinct ways. Emerging science interventions include: Virtual worlds (Bautista and Boone 2015); Cognitive-apprenticeship based instruction (Cooper 2015); Community Links (Yang et al. 2014); and increasingly deep practical science teaching experiences (Cartwright and Atwood 2014; Flores 2015). Bautista and Boone (2015) found that participation in a mixed reality learning environment covaried with significant increases in the PSTE and STOE scores of 62 pre-service teachers. Controlled pedagogical mastery, emotional arousal and self-modelling were identified as contributing factors that would otherwise be unavailable in more traditional science teaching approaches. Yang et al. (2014) found that pre-service teachers showed a large effect size gain in their PSTE scores after completing a content and pedagogy based STEM course. The researchers and participants attributed these gains to the opportunities for service learning afforded by two community partner organisations. A total of 46 pieces of STEBI-B research used a quasi-experimental design with a single cohort. After these studies were removed, 32 remained eligible in the next step of the analysis.

Quasi Experimental Designs with Multiple Cohorts (2.5)

Quasi-experimental research with multiple cohorts affords researchers with unique opportunities to assess interventions across different iterations over time. Ford and others (2012) collected data from three separate cohorts from 2006 to 2008, to assess the relationship between pre-service teachers' participation in science courses focusing on inquiry-learning and problem based learning, and their STEBs. The science course addressed three key science content areas (Physical Science, Biology and Earth Science) in conjunction with science curriculum through inquiry questions, assessment tasks and guided laboratories. Two of the three cohorts had strong PSTE growth. The 2007 cohort showed small to moderate PSTE growth (Cohen's d = 0.3). This finding is in considerable contrast to the 2006 (Cohen's d = 0.94) and 2008 (Cohen's d = 1.1). This difference between the cohorts was not addressed by the authors. Conversely, none of the cohorts showed any significant change in their outcome expectancies. There were no trends over time within this study. Morrell and Carroll (2003) found that the fourth iteration of an inquiry science methods course lead to much higher growth in the personal science teaching efficacy of participants than the previous three. The science and mathematics methods course used extended field experiences (12 h per week) to provide students with opportunities to teach science lessons which they had developed. The PSTE effect size reported in 1997, 1997 and 1999 were 0.206, 0.338 and 0.2 respectively. For the science course offering in 2000, the reported PSTE effect size rose to 0.95. This raises a 'why' question that needs to be answered. Sasser (2014) used a multiple cohorts design to analyse Problem-Based Learning (PBL) with unparalleled depth. Rather than retracing previous trails by assessing the educational impact of PBL, Sasser (2014) explored the structure required for effective implementation. Two cohorts of pre-service teachers were given the same problem-based learning scenario. One cohort was given structural support, whereas the other received no support as they engaged in an open-ended experience. Curiously, there was no significant difference in the science teaching efficacy beliefs between the cohorts. The additional structure did seem to help students to increase their science content knowledge.

An opportunity is being missed with the use of quasi-experimental research designs using multiple cohorts across multiple iterations of a science subject. The increased sample sizes and repeated STEBI-B administrations strengthen the argument for covariance between the key variables, but deeper narratives can be explored. Certainly, a focus on how the changes that are made to science interventions and the subsequent effects of those changes on the STEBs of pre-service teachers represents a deeper, fresher path for future research in this area. There were 12 studies that supplemented quasi experimental with repeated implementations across multiple cohorts. After these studies were removed, 20 were assessed in the next stage of analysis.

Longitudinal Quasi-Experimental Research Designs (3)

Much of the research coded at this level assesses the durability of STEB gains made during a pre- and post-test, quasi-experimental investigation (Ginns et al. 1995; Hechter 2008; Palmer 2006a, b; Richardson and Liang 2008). Palmer (2006a) found the considerable STEB improvements that students experienced as they participated in an innovative science methods course (modelling, inquiry, cooperative learning) remained durable for up to nine months after the course had been completed. Opportunities for mastery experiences were crucial to the consolidation of the preservice teachers' STEBs as their practical science teaching experiences provided tangible evidence of their emerging abilities to both engage students and assist them to meet science learning objectives. Richardson and Liang (2008) chose to administer the STEBI-B in the first week of the second science subject to assess the durability of the efficacious changes that occurred within their first science subject. Not only were the STEB changes durable in the absence of the inquiry-learning science subject, the participants displayed small increases during this period. Ginns and others (1995) took a more holistic approach as they examined STEBs in relation to an entire teacher education program. They found that the STEBs of pre-service teachers did not improve as they completed the teacher education program. Hechter (2008) used a longitudinal framework to explore pre-service teachers' reflections on their educational experiences within a science methods course, rather than to investigate durability. After a delay period, the pre-service teachers were asked to respond to the STEBI-B instrument based on how they felt about science teaching prior to undertaking the science methods course. Upon reflection, their retrospective pre-test scores were much lower than the original pre-test scores. Clearly, they valued the science methods subject after experiencing it in full. There were 8 research articles remaining after the 12 longitudinal pieces of research were removed.

Experimental Research Designs (4)

The remaining 8 research articles used experimental research designs. Scharmann and Orth Hampton (1995) used a robust experimental design with two cohort groups to assess the impact of a science methods course involving hands-on investigation and cooperative learning. The heterogeneous cooperative learning groups did not show higher science teaching efficacy beliefs than those in the control group. McDonnough and Matkins (2010) created an experimental design, strengthened by repeated measures, by collecting data from different institutions over two years. Despite statistical outliers, the results indicated that imbedding science into practical teaching experiences causes larger PSTE increases. Logerwell (2009) showed that problem based learning strategies represent a viable way of increasing pre-service teachers' outcome expectancy beliefs. The study employed two control groups and an experimental group over a 2-week summer science teaching experience.

In a creative solution to ethical issues at the tertiary level, Ebrahim (2012) used a cohort of pre-service teachers enrolled in a practical placement course, with no science component, as a control group. Those who participated in the science methods course displayed moderate STEB growth, whereas the control group showed no STEB change. Thus, the researcher can make the claim that the curriculum design and science teaching experiences caused the increased science-teaching efficacy reported by participants. A similar science methods course showed increased STEBs in the experimental group (Bhattacharyya et al. 2009). Conversely, the control group showed small declines. However, the generalisability of the research is limited by both the small sample size and the lack of subscale differentiation. The following section will describe the use of the PSTE and STOE subscales within the STEBI-B literature.

The PSTE and STOE Subscales Within the STEBI-B Literature

There were 117 articles that provided sufficient information to allow for the subscale use to be analysed. Within the selected articles, there is some inconsistency amongst the usage of the PSTE and STOE subscales, despite the conceptual separation of both subscales (Bleicher 2004; Enochs and Riggs 1990). Simply blending both constructs together does not accommodate the complexity of the targeted constructs and yet 16 pieces of research have done just that. This blending can take the form of merged STEB scores (e.g. Kahraman et al. 2014) or single item analyses (e.g. Urban-Woldron 2014). Such errors may be more prominent in cross-disciplinary educational comparisons where the researchers are perhaps not as familiar with the STEBI-B instrument (Bursal and Paznokas 2006; Saçkes et al. 2012). Conversely, ignorance cannot be blamed in research where the subscales are not differentiated within the results after the author(s) describe them earlier in their writing (e.g. Slater et al. 2008).

Discounting the research with blended subscales, nearly a quarter of all analysed papers did not measure the STOE subscale. Of the 104 papers that formally measured the PSTE subscale, 14 of these did not measure the STOE subscale. The choice to ignore the STOE subscale in favour of the PSTE subscale is becoming more prominent as time passes with 79 % of the research in this category being published after 2007. The implications of the decline in STOE usage will be unpacked in the discussion.

In most studies the PSTE scores of the participants were greater than their STOE scores on all occasions of testing. In total, there were 83 research articles that clearly presented comparable data for both subscales on at least one testing occasion. The mean scores of the PSTE were higher than the STOE on all testing occasions in 92.7 % of these papers. Thus, only three studies exist where the STOE was recorded as greater than the PSTE at any point (e.g. Bayraktar 2011). This trend implies that despite feeling confident in their own abilities, many preservice teachers are not as certain about the effectiveness of science teaching in general. This is unpacked in greater detail in the discussion section.

It appears harder to produce growth within the STOE subscale in comparison to the PSTE subscale. There were 58 papers that allowed for growth comparisons between the subscales because they met the following conditions; the STEBI-B was used at least twice; and the appropriate descriptive statistics were presented clearly. 84.5 % of these papers showed higher growth on the PSTE subscale. Nevertheless, there is some evidence of positive change emerging from the body of literature as 6 of the 7 research items that display greater STOE growth were published in 2009 and beyond. Hopefully, this is a sign of development stemming from reflection upon earlier research rather than an anomaly. The next section will explore the innovative practices used within science interventions.

Innovative Practices Within the Science Interventions

A total of 91 STEBI-B articles included a science intervention as part of the research design. There were 8 articles which did not describe the intervention in sufficient detail for the dichotomous coding of innovative practices. Each of the remaining 83 articles was coded as either employing or not employing each of the 14 identified innovative practices. Figure 1.3 presents the number of research items that used each of the innovative practices. Non Sequenced Content was coded to reflect a more traditional approach to science content course design. There is strong variation in the innovative practices employed within analysed science interventions. The innovative practices are not mutually exclusive of one another and in many instances multiple practices have been amalgamated into complex science education designs.

The most common pedagogical inclusions were curriculum development (43.4 %), inquiry learning (51.8 %) and in-subject practical experience (43.4 %). The prominence of these approaches suggests that the interventions are being thoughtfully designed to suit the purpose of pre-service science education (i.e. producing elementary science teachers). Unsurprisingly, constructivism (34.9 %)

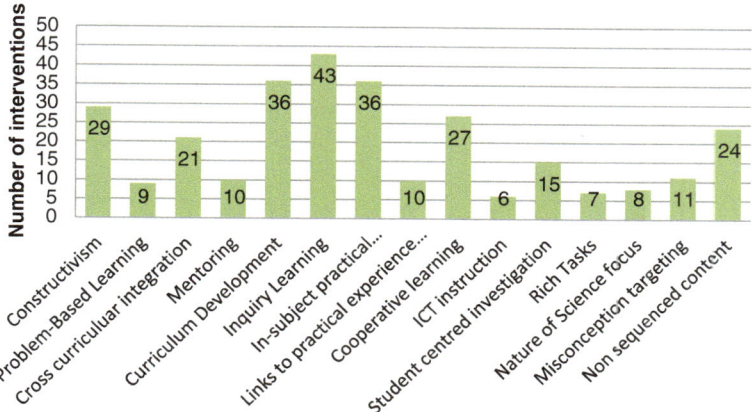

Fig. 1.3 Innovative practices included within the STEBI-B literature

was also cited frequently within the literature. However, despite constructivism being mentioned frequently as an underlying principle it is seldom described in an actionable way. Simply put, the readers need to know how opportunities for constructivist learning have been provided within intervention descriptions. It is not uncommon for educational concepts, such as constructivism, to be broadly outlined without supporting information relating to pedagogical structure (e.g. Plourde 2002). This has led to the researcher to conclude that constructivism is primarily being included in a shallow, tokenistic fashion. This interpretation is supported by the lack of detail and scaffolding that is often evident in cooperative learning inclusions. This is certainly not a criticism of researchers, many of whom are responding to the constraints of their chosen mediums. More broadly speaking, the requirement for detailed pedagogical descriptions represents a need for a holistic shift in the focus of science education research to processes/interventions in combination with findings.

There appear to be themes within the interventions that could be construed as problematic. Firstly, the delivery of varying science concepts on a weekly basis was a frequent theme in this analysis (28.9 %). Such isolated, content focused learning experiences conflict directly with the more integrated, student centred and profession focused interventions that covary with positive outcomes for students. However, it should be noted that the 24 interventions employing the weekly content change strategy generally have supplementary innovative pedagogies in place. Weekly content change generally involves new areas of content focus each week. Week one may focus on biology, week two may focus on geology, week three may focus on chemistry and so forth. It may be challenging, although not impossible, to make rich connections between different content areas in a single semester. Ford and others (2012) were able to overcome the issues of this approach by limiting the semester to four content areas which linked with ongoing inquiry and problem-based learning approaches. The mean number of innovations of this group (3.16) is almost the same as the entire group of analysed interventions (3.23). Secondly, ICT instruction (6 %), rich tasks (11 %) and mentoring (12.6 %) are underrepresented within the literature. The absence of mentoring is particularly disconcerting as this may represent a divide between pre-service and in-service teachers. Such a divide could diminish the positive long term effects of tertiary science education programs. The following section will explore the PSTE and STOE effect sizes reportedly produced by these science interventions.

The PSTE and STOE Effect Sizes Produced by Science Interventions

Prior to analysing the effect sizes for the PSTE and STOE subscales, the studies with less than 21 participants were removed from the analysis. This prevents the potentially inaccurate skewing of data and should allow for a relatively normal distribution of STEBI-B scores within the included research items. Figure 1.4 below shows the distribution of effect sizes reported on both the PSTE and STOE subscales. The red lines show the insignificant, small, moderate, large and very large effect size ranges. It should be noted that despite a slight negative skew, the STOE effect sizes are also normally distributed. The PSTE scores are generally higher with a positive skew as all but one of the very large ES gains were reported on this subscale. The Kurtosis scores of the PSTE and STOE effect size data sets show further subscale differences. The PSTE (-0.272) Kurtosis is close to zero, suggesting a relatively normal distribution curve. In comparison, the STOE Kurtosis (0.723) shows a flatter distribution of scores spread further from the mean. This would appear to reflect both the inconsistent measurement and frequent stagnation of the STOE subscale.

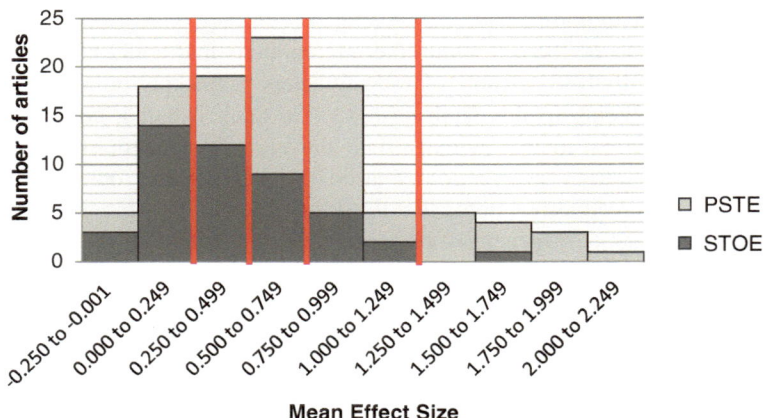

Fig. 1.4 PSTE and STOE distribution histogram

Statistical analysis indicates that there is a substantial difference between the PSTE and STOE subscales in terms of mean effect size produced within the body of literature. Table 1.3 presents the descriptive statistics for the PSTE and STOE effect sizes. The mean effect size produced on the STOE subscale is moderate (0.43) and only approximately half of that shown on the PSTE subscale (0.83). This trend is representative of the wider body of STEBI-B literature as PSTE growth is almost always higher than STOE growth. This is evident in 84.5 % of relevant cases. There were 9 research items included in the effect size analyses which did not measure the STOE subscale, despite correctly utilising the PSTE subscale. There may be lower effect sizes on the subscale that are not being reported within the literature.

Even though there are substantial statistical differences between the mean scores and effects sizes on both the PSTE and STOE subscales, a statistically significant correlation exists between these science efficacy measures. Table 1.4 shows the output from the correlation analysis conducted on the mean PSTE and STOE effect sizes. The correlation analysis shows that there is a statistically significant moderate-to-strong correlation (Pearson's R = 0.628) between the PSTE and STOE effect sizes. These findings indicate that the STOE needs to be considered alongside PSTE rather than dismissed for science teacher education. The issues with the STOE subscale will be unpacked further on in the discussion section of this chapter. The following paragraphs will rank the PSTE and STOE effect sizes within the literature and unpack the pedagogical themes.

The variation in innovative practices employed within the top science interventions in terms of PSTE effect sizes indicates that there is no 'simple' solution to improving the science outcomes of pre-service elementary teachers. Table 1.5 ranks the top research pieces on PSTE effect size changes and lists the identified innovative practices. The author recognises that the innovative practices listed may be limited by the framework. It is advised that the reader refer to the original articles for more accurate information. The most recurrent innovative

Table 1.3 Descriptive statistics for PSTE and STOE effect sizes

	Mean	Std. Deviation	N
MeanPSTE	0.833	0.512	55
MeanSTOE	0.429	0.357	46

Table 1.4 Correlation analysis for PSTE and STOE effect sizes

		MeanPSTE	MeanSTOE
MeanPSTE	Pearson correlation	1	0.628[**]
	Sig. (2-tailed)		0.000
	N	55	46
MeanSTOE	Pearson correlation	0.628[**]	1
	Sig. (2-tailed)	0.000	
	N	46	46

[**]Correlation is significant at the 0.01 level (2-tailed)

Table 1.5 Top 10 PSTE effect sizes

Year	Author(s)	Innovative practices	Effect size
2002	Jabot	Curriculum development, links to professional experience blocks and alternative conception targeting	2.0
2015	Cooper	Mentoring, curriculum development, in-subject practical experience	1.93
2006a	Palmer	Cooperative learning, student-centred investigation	1.87
2011	Bautista	Integration with other KLAs, inquiry learning, in-subject practical experience, link to professional experience blocks, nature of science focus, alternative conception targeting	1.83
2010	Swars and Dooley	Mentoring, curriculum development, inquiry learning, in-subject practical experience, links to professional experience blocks, student-centred investigation	1.67
2015	Flores	Curriculum development, link to professional experience blocks, cooperative learing	1.59
2009	Bleicher	Constructivism, curriculum development, cooperative learning	1.58
2009	Logerwell	Problem-based learning. curriculum development, inquiry learning, in-subject practical experience	1.44
2014	Yang et al.	Constructivism, integration with other KLAs, inquiry learning, in-subject practical experience, cooperative learning	1.3
2012	Brower	In-subject practical experience	1.27

practices amongst these 10 research items were in-subject practical experience (6), inquiry learning (5) and curriculum development (5). Of interest was the limited integration of misconception targeting (1) and nature of science teaching (1) amongst the PSTE top 10. The mean number of interventions within this group (3.6) was slightly larger than the mean produced by the entire group of analysed interventions (3.2).

Jabot's (2002) science intervention may be a viable solution to the 2-Sigma problem (Bloom 1984) in relation to the PSTE of pre-service elementary teachers. A total of 24 pre-service teachers participated in a reflective, misconception based intervention. The students were required to develop a science unit of work that aimed to redress specific misconceptions held by elementary students. The intervention culminated in a 'Teaching Participation' practical experience where the pre-service teachers implemented their units of work to students of the appropriate age level. More generally, practical science teaching experiences seem to be related to larger PSTE gains (Bautista 2011; Brower 2012; Cantrell 2003; Logerwell 2009).

Complex science interventions with multiple innovative practices covary positively with growth on the PSTE scale (Bautista 2011). The misconception targeting and practical experience elements were supplemented with opportunities for the participants to observe the science teaching of accomplished teachers. Science teaching dvds, practical experience, science teaching observations and tutor modelling were used to deliver vicarious learning opportunities to pre-service primary teachers. This type of sophisticated pedagogical design would require strong inter-faculty relationships within teacher education programs and collaborative partnerships with elementary schools. Indeed, broad reform at a program level covaries with the improved science teaching efficacy beliefs of pre-service teachers (Wenner 1995). Given the declining state of elementary science education, in nations such as Australia, stemming from diminished curriculum time (Angus et al. 2004; Goodrum et al. 2001; Goodrum and Rennie 2007; Tytler 2007; Tytler et al. 2009) it would be challenging to implement similar vicarious experiences within the Australian tertiary context. Nevertheless, vicarious learning experiences could be used to improve stagnant STOEs. The following paragraphs rank and discuss the stronger research in terms of STOE outcomes for participants

Many of the interventions, that produced high STOE effect sizes, showed deep pedagogical consideration through the use of multiple innovative practices. In fact, the top ranked STOE science interventions employed an average of 4.6 innovative practices. Table 1.6 ranks the top 10 research items on STOE effect size. Amongst these research items, cooperative learning (6), professional relevance (6), inquiry learning (6) and in-subject practical experience (6) were the most frequently used innovative practices. The biggest change from the PSTE items was the emergence of both cooperative learning approaches (6) and student centred investigation (4). An interpretation of this could be that cooperative learning extends a participant's focus beyond the immediate self by allowing for meaningful collaboration with other prospective teachers, thus affording the necessary broader experiential learning necessary to affect change on the STOE subscale.

Curiously, two of the highest STOE performing articles chose to 'simplify' their educational designs to allow for a deep implementation of the chosen innovations (Ozdelik and Bulunuz 2009; Palmer 2006a). Ozdelik and Bulunuz (2009) reported on a traditional 'weekly topics', with content variations, approach to tertiary science education. The different content areas were supplemented with inquiry-based background research and 'hands-on' investigations. Palmer (2006a) employed a similar approach with the student-centred delivery of content pitched at the elementary level. Although less 'academically rigorous' the STEBI-B data indicates that these approaches help to alleviate the effects of pre-service teachers' detrimental science experiences and negative attitudes to science by presenting the subject in a more accessible, engaging and professionally appropriate manner.

Deeply drilled, complementary science subjects within a tertiary education program can produce significant and durable growth in the STOEs of pre-service elementary teachers (Cross 2010; Deehan 2013). The first year science subject used Astronomy content to drive a misconception-based, inquiry based approach where the pre-service teachers were required to develop their Pedagogical Content Knowledge (Cross 2010). The second science subject implemented a collaborative, PBL scenario that afforded the pre-service teachers the opportunity to engage in a professional environment while developing their knowledge of the

Table 1.6 Top 10 STOE effect sizes

Year	Author(s)	Innovative practices	Effect size
2002	Jabot	Curriculum development, links to professional experience blocks and alternative conception targeting	1.54
2015	Cooper	Mentoring, curriculum development, in-subject practical experience	1.13
2009	Ozdelik and Bulunuz	Inquiry learning, cooperative learning	1.11
2006a, b	Palmer	Cooperative learning, student-centred investigation	0.92
2011	Bautista	Integration with other KLAs, inquiry learning, in-subject practical experience, link to professional experience blocks, nature of science focus, alternative conception targeting	0.85
2010	Cross	Constructivism, mentoring, curriculum development, inquiry learning, in-subject practical experience, cooperative learning, student-centred investigation, alternative conception targetting	0.81
2009	Bleicher	Constructivism, curriculum development, cooperative learning	0.75
1999	Wingfield and Ramsey	Mentoring, curriculum development, in-subject practical experience, cooperative learning	0.71
2013	Deehan	Constructivism, problem-based learning, Integration with other KLAs, curriculum development, inquiry learning, in-subject practical experience, cooperative learning, ICT instruction. student-centred investigation, Rich Tasks	0.70
2015	Knaggs and Sondergeld	Curriculum development, inquiry learning, in-subject practical experience, student-centred investigation, nature of science instruction	0.69

elementary science curriculum (Deehan 2013). Both subjects produced STOE effect sizes that were significantly higher than the literature mean and were durable for up to 12-months after the end of the second subject. Also, it should be noted that this cohort of pre-service teachers displayed equivalent mean scores on both the PSTE and the STOE subscale on the final STEBI-B administration. Such subscale equality is unprecedented within the STEBI-B literature and represents an ideal goal in the science teaching efficacy of pre-service elementary teachers. However, given the lack of research linking pre-service to in-service teaching, it remains unclear how such efficacious gains influence the science teaching practices of in-service teachers. The following section will discuss the implications and research directions that should arise from the STEBI-B analyses conducted within this chapter.

Discussion

Throughout its 25 year existence, the STEBI-B has been employed in a variety of compelling and worthwhile research projects. The instrument has been used as: a basis for deep qualitative interviews (e.g. Tosun 2000); a means of assessing science teaching efficacy beliefs with other variables (e.g. Serin and Bayraktar 2014); a way of assessing science education subjects (e.g. Swars and Dooley 2010); and even as; a way of assessing the cumulative effects of entire teacher education programs (e.g. Ginns et al. 1995). The reliability and validity of the STEBI-B instrument has lead to its use as a basis for the development of alternate efficacy instruments (e.g. Wilson 2012). The body of literature appears to show a shift from identifying science teaching efficacy issues to rectifying such issues. This is evident as there has been a trend towards rigorous research designs with multiple administrations of the STEBI-B instrument. In fact, multiple administrations of the STEBI-B instrument were used to obtain relevant statistical data in 60 % of the analysed articles. Yet, pathways remain for the improvement of the methodological implementation of the STEBI-B instrument within the body of literature. The overwhelming absence of experimental designs in the literature prevents the establishment of causal relationships between science interventions and STEBs. Although, many researchers have attempted to strengthen the argument for covariance by reporting on multiple cohorts (e.g. Cone 2009; Ford et al. 2011), there are still only 8 research items that have employed the STEBI-B in an experimental framework. The widespread lack of causal relationships within the STEBI-B literature may, at least in some part, be connected to the large variation in the pedagogical practices employed within tertiary science interventions (e.g Palmer 2007).

The STOE subscale is problematic in comparison with the PSTE subscale. The body of literature reveals that the STOE is almost always lower than the PSTE on all administrations of the STEBI-B. In fact, of the 83 articles that measured both the PSTE and STOE subscales on at least on occasion, the mean STOE score was higher than the mean PSTE score only five times (e.g. Bayraktar 2011; Cross 2010; Deehan 2013). There is a similar disparity in the effect size changes reported on the STEBI-B subscales. The mean effect size change on the PSTE scale (0.83) is nearly twice as large as that on the STOE scale (0.43). In summation, pre-service primary teachers generally feel more efficacious about their capacity to teach science effectively then they feel about the ability of science teachers in general to guide students towards desired learning outcomes.

Criticisms of both the validity and reliability of the STOE subscale have been presented as reasons for its removal from research designs (Andersen et al. 2004; Bursal 2008; Cannon and Scharmann 1996; McDonnough and Matkins 2010; Velthuis et al. 2014). Bursal (2010) diminished the validity of the construct by claiming that the statements comprising the STOE subscale align with a 'teacher-centred' approach to science teaching that does not reflect modern educational principles. An analysis of the STOE items refutes this interpretation. Indeed, the statements comprising the STOE subscale appear to be pedagogically neutral.

Other researchers cite the low reliability of the STOE measure as a reason for dismissal (Andersen et al. 2004; Velthuis et al. 2014). Velthuis and others (2014) found that the STOE subscale produced a Cronbach's alpha of 0.56 in a Danish university and subsequently removed it from the research. Such findings are common within the literature as the reliability of the STOE subscale is generally lower than the PSTE subscale (Aydin and Boz 2010; Bleicher 2004; Cross 2010; Deehan 2013; Enochs and Riggs 1990). Cannon and Scharmann (1996) appear to be resigned to low reliability on the STOE subscale, as they believe that pre-service teachers lack the necessary conceptualisations of the teaching profession to respond to the STOE statements appropriately. It could be argued that it is the purpose of pre-service teacher training to provide students with the opportunities to develop such conceptual understandings of the profession. Longitudinal research reveals that the reliability of the STOE subscale improves as pre-service teachers progress through their degrees, even in the absence of formal science education (Cross 2010; Deehan 2013).

There is considerable pedagogical variation amongst the science interventions presented in the STEBI-B literature. Curriculum development, inquiry learning, and in-subject practical experiences are the most common pedagogical inclusions within the STEBI-B literature. ICT instruction, links to professional experiences placements and problem-based learning were all conspicuously absent from the body of research. Educational designs with multiple innovative practices and deep collaboration beyond the immediate subject tend to covary with higher effect sizes on the PSTE and STOE subscales. Student centred approaches and practical science teaching experiences were used within the science interventions that produced the strongest growth in personal science teaching efficacy. Conversely, analyses revealed that there is no simple pedagogical solution to producing high effect size gains on the STOE subscale. This can likely can attributed to the varied, external locus of control of the broader science teaching outcome expectancy subscale. The number of innovations used within science interventions appears to be a stronger predictor of STOE growth rather than the types of innovations used. Yet, the simplification of content also covaries with improved STOE scores (Ozdelik and Bulunuz 2009; Palmer 2006a, b). Such dissonance between content and pedagogies in the research trends resembles the broader issues that are frequently mentioned in relation to the STOE subscale.

While many researchers are reporting positive correlations between science interventions and the STEBs of participants, the durability of any positive changes remain unknown. Only 8.5 % of the analysed research items assessed the durability of the participants' STEB changes in the absence of a formal science treatment. If the purpose of the tertiary science education programs is to prepare future teachers to deliver quality science education, then logic dictates that the durability of intervention outcomes must be considered both within and beyond the tertiary context. Currently there are few articles that extend the STEBI-B literature into the in-service teaching domain (e.g. McKinnon and Lamberts 2013). Given the changing demographics of the elementary teaching workforce, as the 'baby boomer' generation nears retirement (Harris and Farrell 2007; NSW DEC 2011), it is imperative that the transition from pre-service to in-service teaching becomes a major research focus in the future.

The framework developed for the organisation of the STEBI-B literature serves the primary function of organising a large body of literature into a coherent format whilst still allowing for broader trends to be analysed. The methodological funnel allows the reader to consider the different methodological and content contributions made to the science teaching efficacy field. Each level of the funnel establishes knowledge that is built upon in the levels that follow. The framework has allowed for innovative practices to be identified and for science teaching efficacy belief effect sizes to be compared across different contexts. Nevertheless, we are still confronted by the "question mark" showing that a relationship exists between reported innovative practices and science teaching efficacy effect size. As an outsider, the researcher cannot know what complex interactions occurred across each classroom within each university to lead to the reported changes. Even still, we cannot yet know what these "changes" mean in a tangible fashion. This provides an argument for the mixed methods approach to science education research. It should be noted, that while not an explicit focus of this review paper, many researchers recognised these issues and employed mixed methods designs (e.g. Bleicher and Lindgren 2005; Leonard et al. 2011; Scott 2013).

The findings presented within this STEBI-B review have considerable implications for the direction of further research. Firstly, the STEBI-B needs to be adopted in contexts beyond Australia, Turkey and the USA. Specifically, more research into the reliability of the STEBI-B subscales needs to be conducted beyond the USA. Currently, there is a tendency to restate the reliabilities reported by the seminal authors (Enochs and Riggs 1990) and other major updates (Bleicher 2004). Secondly, the STEBI-B should be used with a greater number of longitudinal and experimental research designs. The use of these more complex research designs would serve the dual purposes of allowing for causal links between tertiary science interventions and reported STEB changes. Longitudinal follow-ups can assess the durability of efficacious gains beyond tertiary contexts. Presently, the STEBI-B literature exemplifies the disconnection between research into the pre-service and in-service domains of elementary science education. More research needs to employ the STEBI-B and STEBI-A instruments to traverse the gap between pre-service and in-service teaching to determine if STEBs remain durable after teachers leave the tertiary context. Thirdly, more researchers should consider presenting the narrative of subject development over time. Given that the most successful science interventions feature complex pedagogical structures, overviews as to how these interventions were developed would serve as meaningful models for replication. Finally, the STOE subscales needs to be considered in both the development of science interventions and the presentation of research. This review has shown that the outcome expectancies of pre-service teachers can be improved with pedagogically complex, student-centred science interventions. While the arguments for the dismissal of the STOE subscale are compelling, addressing these issues would advance the body of research into valuable new directions.

References

Andersen, A. M., Dragsted, S., Evans, R. H., & Sørensen, H. (2004). The relationship between changes in teachers' self-efficacy beliefs and the science teaching environment of Danish first-year elementary teachers. *Journal of Science Teacher Education, 15*(1), 25–38.

Angus, M., Onley, H., Ainley, J., Caldwell, B., Burke, G., & Selleck, R. (2004). *The sufficiency of resources for Australian primary schools.* Canberra: Commonwealth Department of Education, Science and Training.

Arigbabu, A. A., & Oludipe, D. I. (2010). Perceived efficacy beliefs of prospective Nigerian Science teachers. *Journal of Science Education and Technology, 19*(1), 27–31.

Avery, L. M., & Meyer, D. Z. (2012). Teaching science as science is practiced: Opportunities and limits for enhancing preservice elementary teachers' self-efficacy for science and science teaching. *School Science and Mathematics, 112*(7), 395–409.

Aydin, S., & Boz, Y. (2010). Pre-service elementary science teachers' science teaching efficacy beliefs and their sources. *Elementary Education Online, 9*(2), 694–704.

Azer, S. A. (2001). Problem-based learning. *Saudi Medical Journal, 22*(4), 299–305.

Bahcivan, E., & Kapucu, S. (2014). Turkish preservice elementary science teachers' conceptions of learning science and science teaching efficacy beliefs: Is there a relationship? *International Journal of Environmental and Science Education, 9*(4), 429–442.

Bautista, N. U. (2011). Investigating the use of vicarious and mastery experiences in influencing early childhood education majors' self-efficacy beliefs. *Journal of Science Teacher Education, 22*(4), 333–349.

Bautista, N. U., & Boone, W. J. (2015). Exploring the impact of TeachME™ lab virtual classroom teaching simulation on early childhood education majors' self-efficacy beliefs. *Journal of Science Teacher Education, 26*(3), 237–262.

Bayraktar, S. (2011). Turkish preservice primary school teachers' science teaching efficacy beliefs and attitudes toward science: The effect of a primary teacher education program. *School Science and Mathematics, 111*(3), 83–92.

Bhattacharyya, S., Volk, T., & Lumpe, A. (2009). The influence of an extensive inquiry-based field experience on pre-service elementary student teachers' science teaching beliefs. *Journal of Science Teacher Education, 20*(3), 199–218.

Bleicher, R. E. (2004). Revisiting the STEBI-B: Measuring self-efficacy in preservice elementary teachers. *School Science and Mathematics, 104*(8), 383–391.

Bleicher, R. E. (2009). Variable relationships among different science learners in elementary science-methods courses. *International Journal of Science and Mathematics Education, 7*(2), 293–313.

Bleicher, R. E., & Lindgren, J. (2005). Success in science learning and preservice science teaching self-efficacy. *Journal of Science Teacher Education, 16*(3), 205–225.

Bloom, B. S. (1984). The 2 sigma problem: The search for methods of group instruction as effective as one-to-one tutoring. *Educational Researcher*, 4–16.

Brower, D. J. (2012). *Incorporating formative assessment and science content into elementary science methods* (Doctoral dissertation, Montana State University-Bozeman, College of Education, Health and Human Development).

Burns, R. (2000). *Introduction to research methods* (3rd ed.). Melbourne: Longman Publishers.

Bursal, M. (2008). Changes in Turkish pre-service elementary teachers' personal science teaching efficacy beliefs and science anxieties during a science method course. *Journal of Turkish Science Education, 5*(1), 99–112.

Bursal, M. (2010). Turkish preservice elementary teachers' self-efficacy beliefs regarding mathematics and science teaching. *International Journal of Science and Mathematics Education, 8*(4), 649–666.

Bursal, M., & Paznokas, L. (2006). Mathematics anxiety and preservice elementary teachers' confidence to teach mathematics and science. *School Science and Mathematics, 106*(4), 173–180.

Çakiroglu, J., Çakiroglu, E., & Boone, W. (2005). Pre-service teacher self-efficacy beliefs regarding science teaching: A comparison of pre-service teachers in Turkey and the USA. *Science Educator, (74)1*, 31–41.

Cannon, J. R., & Scharmann, L. C. (1996). Influence of a cooperative early field experience on preservice elementary teachers' science self-efficacy. *Science Education, 80*(4), 419–436.

Cantrell, P. (2003). Traditional versus retrospective pretests for measuring science teaching efficacy beliefs in preservice teachers. *School Science and Mathematics, 103*(4), 177–185.

Cartwright, T. J., & Atwood, J. (2014). Elementary pre-service teachers' response-shift bias: Self-efficacy and attitudes toward science. *International Journal of Science Education, 36*(14), 2421–2437.

Cone, N. (2009). Community-based service-learning as a source of personal self-efficacy: Preparing preservice elementary teachers to teach science for diversity. *School Science and Mathematics, 109*(1), 20–30.

Cooper, T. O. (2015). *Investigating the effects of cognitive apprenticeship-based instructional coaching on science teaching efficacy beliefs* (Doctoral dissertation, Florida International University).

Cross, M. (2010). *The journeys of pre service primary teachers in learning how to teach science: A longitudinal case study*. Charles Sturt University, Bathurst, 1–157. Unpublished Honours Thesis, Charles Sturt University.

Deehan, M. (2013). *How do I measure up? A longitudinal investigation of a cohort of Australian pre-service primary teachers' science experiences*. Unpublished Honours Thesis. Charles Sturt University, Bathurst, 1–190.

Duran, E., Ballone-Duran, L., Haney, J., & Beltyukova, S. (2009). The impact of a professional development program integrating informal science education on early childhood teachers' self-efficacy and beliefs about inquiry-based science teaching. *Journal of Elementary Science Education, 21*(4), 53–70.

Ebrahim, A. H. (2012). The self-efficacy of preservice elementary teachers in Kuwaiti science programs. *Education, 133*(1), 67–76.

Edelson, D. C., Gordin, D. N., & Pea, R. D. (1999). Addressing the challenges of inquiry-based learning through technology and curriculum design. *Journal of the Learning Sciences, 8*(3–4), 391–450.

Englehart, D. (2008). *An exploration of how pre-service early childhood teachers use educative curriculum materials to support their science teaching practices* (Doctoral dissertation, University of Central Florida Orlando, Florida).

Enochs, L. G., & Riggs, I. M. (1990). Further development of an elementary science teaching efficacy belief instrument: A preservice elementary scale. *School Science and Mathematics, 90*(8), 694–706.

Flores, I. M. (2015). Developing preservice teachers' self-efficacy through field-based science teaching practice with elementary students. *Research in Higher Education, 27*(1), 1–20.

Ford, D. J., Allen, D., Dagher, Z., & Donham, R. (2011). *Reforming science and science education courses for K-8 pre-service teachers: The University of Delaware teacher professional continuum project*. Paper presented at the NSEUS National Conference on Research Based Undergraduate Science Teaching: Investigating Reform in Classrooms, Bryant Conference Center, University of Alabama, Tuscaloosa AL.

Ford, D. J., Fifield, S., Madsen, J., & Qian, X. (2012). The science semester: Cross-disciplinary inquiry for prospective elementary teachers. *Journal of Science Teacher Education, 24*(6), 1049–1072.

Gencer, A. S., & Çakiroglu, J. (2007). Turkish preservice science teachers' efficacy beliefs regarding science teaching and their beliefs about classroom management. *Teaching and Teacher Education, 23*(5), 664–675.

Ginns, I. S., Tulip, D. F., Watters, J. J., & Lucas, K. B. (1995). Changes in preservice elementary teachers' sense of efficacy in teaching science. *School Science and Mathematics, 95*(8), 394–400.

Goodrum, D., & Rennie, L. (2007). *Australian School Science Education—National Action Plan 2008–2012* (Vol. 1). Canberra: Commonwealth of Australia.

Goodrum, D., Hackling, M., & Rennie, L. (2001). *The status and quality of teaching and learning of science in Australian schools*. Canberra: Department of Education, Training and Youth Affairs.

Green, J. L., Camilli, G., & Elmore, P. B. (Eds.). (2006). *Handbook of complementary methods in education research*. Routledge.

Griffith, G., & Scharmann, L. (2008). Initial impacts of No Child Left Behind on elementary science education. *Journal of Elementary Science Education, 20*(3), 35–48.

Harris, K. L., & Farrell, K. (2007). The science shortfall: An analysis of the shortage of suitably qualified science teachers in Australian schools and the policy implications for universities. *Journal of Higher Education Policy and Management, 29*(2), 159–171.

Hechter, R. (2008). *Changes in preservice elementary teachers' personal science teaching efficacy and science teaching outcome expectancies: The influence of context*. Unpublished Doctoral Dissertation, University of North Dakota.

Hechter, R. (2010). Changes in preservice elementary teachers' personal science teaching efficacy and science teaching outcome expectancies: The influence of context. *Journal of Science Teacher Education, 22*(1), 187–202.

Hudson, P. (2004). Specific mentoring: A theory and model for developing primary science teaching practices. *European journal of teacher education, 27*(2), 139–146.

Ivey, T., Colston, N., & Thomas, J. (2015). Bringing space science down to Earth for preservice elementary teachers. *Electronic Journal of Science Education, 19*(2).

Jabot, M. (2002). The effectiveness of a misconceptions-based approach to the teaching of elementary science methods. *Proceedings of the Pathways to Change International Conference on Transforming Math and Science Education in the K16 Continuum*. Arlington VA: The Science, Technology, Engineering and Mathematics Teacher Education Collaborative (STEMTEC).

Johnston, J. (2003). Active learning and pre service teacher attitudinal change. *Educational Research Association, 3*–23.

Kahn, P., & O'Rourke, K. (2005). Understanding equiry-based learning. In T. Barrett, I. M. Labhrainn, & H. Fallon (Eds.), *Handbook of Enquiry- and Problem- based learning: Irish case studies and international perspectives* (pp. 1–12). Dublin: Centre for Excellence in Learning and Teaching, NUI Galway and All Ireland Society for Higher Education (AISHE).

Kahraman, S., Yilmaz, Z. A., Bayrak, R., & Gunes, K. (2014). Investigation of pre-service science teachers' self-efficacy beliefs of science teaching. *Procedia-Social and Behavioral Sciences, 136*, 501–505.

Kenny, J. (2010). Preparing pre-service primary teachers to teach primary science: A partnership-based approach. *International Journal of Science Education, 32*(10), 1267–1288.

Kervin, L., Wilma, V., Herrington, J., & Okely, T. (2006). *Research for educators*. Melbourne: Cengage learning Australia.

King, K. P., & Wiseman, D. L. (2001). Comparing science efficacy beliefs of elementary education majors in integrated and non-integrated teacher education coursework. *Journal of Science Teacher Education, 12*(2), 143–153.

Knaggs, C. M., & Sondergeld, T. A. (2015). Science as a learner and as a teacher: Measuring science self-efficacy of elementary preservice teachers. *School Science and Mathematics, 115*(3), 117–128.

Lawrance, G. A., & Palmer, D. A. (2003). *Clever teachers, clever sciences: Preparing teachers for the challenge of teaching science, mathematics and technology in 21st Century Australia*. Canberra: Australian Government, Department of Education, Science and Training: Research Analysis and Evaluation Group.

Leonard, J., Barnes-Johnson, J., Dantley, S. J., & Kimber, C. (2011). Teaching science inquiry in urban contexts: The role of elementary preservice teachers' beliefs. *The Urban Review, 43*(1), 124–150.

Lewthwaite, B., Murray, J., & Hechter, R. (2012). Revising teacher candidates' views of science and self: Can accounts from the history of science help?. *International Journal of Environmental and Science Education, 7*(3).

Logerwell, M. G. (2009). *The effects of a summer science camp teaching experience on preservice elementary teachers' science teaching efficacy, science content knowledge, and understanding of the nature of science* (Doctoral dissertation, George Mason University).

Luera, G. R., & Otto, C. A. (2005). Development and evaluation of an inquiry-based elementary science teacher education program reflecting current reform movements. *Journal of Science Teacher Education, 16*(3), 241–258.

Mashnad, P. F. (2008). *An alternate approach to pre-service teachers' misconceptions about biology subject* (Doctoral dissertation, University of Texas at Dallas).

McDonnough, J. T., & Matkins, J. J. (2010). The role of field experience in elementary preservice teachers' self-efficacy and ability to connect research to practice. *School Science and Mathematics, 110*(1), 13–23.

McKinnon, M., & Lamberts, R. (2013). Influencing science teaching self-efficacy beliefs of primary school teachers: A longitudinal case study. *International Journal of Science Education, Part B*, (ahead-of-print), 1–23.

Morrell, P. D., & Carroll, J. B. (2003). An extended examination of preservice elementary teachers' science teaching self-efficacy. *School Science and Mathematics, 103*(5), 246–251.

Mulholland, J., Dorman, J. P., & Odgers, B. M. (2004). Assessment of science teaching efficacy of preservice teachers in an Australian university. *Journal of Science Teacher Education, 15*(4), 313–331.

Naidoo, K. (2013). Transforming beliefs and practices: Elementary teacher candidates' development through shared authentic teaching and reflection experiences within an innovative science methods course. *Dissertation Abstracts Internal, 74*(10), 256.

Newsome, D. L. (2003). The efficacy beliefs of preservice science teachers in professional development school and traditional school settings. *Dissertation Abstracts International, 64*(5), 1588–1694.

NSW Government—Department of Education and Communities. (2011). *2011 Workforce plan— for school teachers in NSW public schools*. Sydney: NSW Government—Department of Education and Communities.

Olgan, R., Alpaslan, Z. G., & Öztekin, C. (2014). Factors influencing pre-service early childhood teachers' outcome expectancy beliefs regarding science teaching. *Egitim Ve Bilim-Education and Science, 39*(173), 288–298.

Ozdelik, Z., & Bulunuz, N. (2009). The effect of a guided inquiry method on pre-service teachers' science teaching self-efficacy beliefs. *Turkish Science Education Journal, 6*(2), 24–42.

Palmer, D. (2006a). Durability of changes in self-efficacy of preservice primary teachers. *International Journal of Science Education, 28*(6), 655–671.

Palmer, D. (2007). Practices and innovations in Australian science teacher education programs. *Journal of Science Education, 38*(2), 167–188.

Palmer, D. H. (2006b). Sources of self-efficacy in a science methods course for primary teacher education students. *Research in Science Education, 36*(4), 337–353.

Park, S. (1996). *Development and validation of the Korean science teaching efficacy beliefs instrument (K-STEBI) for prospective elementary school teachers* (Doctoral dissertation, Pennsylvania State University).

Peters-Burton, E. E., & Hiller, S. E. (2013). Fun science: The use of variable manipulation to avoid content instruction. *Journal of Science Teacher Education, 24*(1), 199–217.

Plourde, L. A. (2002). The influence of student teaching on pre-service elementary teachers' science self-efficacy and outcome expectancy beliefs. *Journal of Instructional Psychology, 29*(4), 245–253.

Ramey-Gassert, L., Shroyer, M. G., & Staver, J. R. (1996). A qualitative study of factors influencing science teaching self-efficacy of elementary level teachers. *Science Education, 80*, 283–315.

Richardson, G. M., & Liang, L. L. (2008). The use of inquiry in the development of preservice teacher efficacy in mathematics and science. *Journal of Elementary Science Education, 20*(1), 1–16.

Richardson, G. M., Liang, L. L., & Wake, D. G. (2014). Examining the durability of environmental education self-efficacy beliefs in preservice teaching. *Applied Environmental Education & Communication, 13*(1), 38–47.

Ritter, J. M. (1999). *The development and validation of the self-efficacy beliefs about equitable science teaching and learning instrument for prospective elementary teachers* (Doctoral dissertation, The Pennsylvania State University).

Rogers, G. & Watters, J. J. (2002). Global perspectives of science education: Successes and challenges of a pilot project. Paper presented at the Annual Meeting of the Australasian Science Education Research Association Conference, Melbourne, July 9–12, 2002.

Saçkes, M., Flevares, L. M., Gonya, J., & Trundle, K. C. (2012). Preservice early childhood teachers' sense of efficacy for integrating mathematics and science: Impact of a methods course. *Journal of Early Childhood Teacher Education, 33*(4), 349–364.

Sarikaya, H., Çakiroglu, J., & Tekkaya, C. (2005). Self-efficacy, attitude and science knowledge. *Academic Exchange Quarterly, 9*(4), 38–42.

Sasser, S. K. (2014). *Effect of structure in problem based learning on science teaching efficacy beliefs and science content knowledge of elementary preservice teachers.* (Doctoral dissertation, Southern Illinois University).

Scharmann, L. C., & Orth Hampton, C. M. (1995). Cooperative learning and preservice elementary teacher science self-efficacy. *Journal of Science Teacher Education, 6*(3), 125–133.

Schmidt, D. L., Saigo, B. W., & Stepans, J. I. (2006). *Conceptual change model: The CCM handbook.* Saiwood Publications.

Schoon, K. J., & Boone, W. J. (1998). Self-efficacy and alternative conceptions of science of preservice elementary teachers. *Science Education, 82*(5), 553–568.

Scott, C. S. (2013). *Inquiry, efficacy, and science education.* (Doctoral dissertation, Georgia Southern University, United States, Georgia). Retrieved February 3, 2014, from http://digitalcommons.georgiasouthern.edu/etd

Serin, K., & Bayraktar, Ş. (2014). Pre-service classroom teachers' science teaching efficacy beliefs and their locus of control status. In *Conference proceedings. New perspectives in science education* (p. 319). libreriauniversitaria. it Edizioni.

Shroyer, M. G., Wright, E. L., & Ramey-Gassert, L. (1996). An innovative model for collaborative reform in elementary school science teaching. *Journal of Science Teacher Education, 7*(3), 151–168.

Sindel, K. D. (2010). *Can experiential education strategies improve elementary science teache' perceptions of and practices in science teaching?* (Doctoral dissertation, Lindenwood University).

Slater, S. J., Slater, T. F., & Shaner, A. (2008). Impact of backwards faded scaffolding in an astronomy course for pre-service elementary teachers based on inquiry. *Journal of Geoscience Education, 56*(5), 408.

Slavin, R. (1991). Synthesis of research on cooperative learning. *Educational Leadership, 48*(5), 71–82.

Soprano, K., & Yang, L. L. (2013). Inquiring into my science teaching through action research: A case study on one pre-service teacher's inquiry-based science teaching and self-efficacy. *International Journal of Science and Mathematics Education, 11*(6), 1351–1368.

Sunger, M. (2007). *An analysis of efficacy beliefs, epistemological beliefs and attitudes towards science in preservice elementary science teachers and secondary science teachers* (Doctoral dissertation, Middle Easter Technical University).

Suter, W. N. (2006). *Introduction to educational research: A critical thinking approach.* Thousand Oaks: Sage Publications.

Swars, S. L., & Dooley, C. M. (2010). Changes in teaching efficacy during a professional development school-based science methods course. *School Science and Mathematics, 110*(4), 193–202.

Tekkaya, C., Çakiroglu, J., & Ozkan, O. (2004). Turkish pre-service science teachers' understanding of science and their confidence in teaching it. *Journal of Education for Teaching, 30*(1), 57–68.

Templeton, C. K. (2007). *The impact of a museum-based science methods course on early childhood/elementary pre-service teachers' self-efficacy and ability to develop curriculum using a constructivist approach*. Florida Atlantic University.

Ting, W., & Albion, P. R. (2014). Remote access laboratories enhancing STEM education. In *Proceedings of the 31st Annual Conference of the Australasian Society for Computers in Learning in Tertiary Education (ASCILITE 2014)* (pp. 579–583). Macquarie University.

Tosun, T. (2000). The beliefs of preservice elementary teachers toward science and science teaching. *School Science and Mathematics, 100*(7), 374–379.

Tytler, R. (2007). *Re-imagining science education: Engaging students in science for Australia's future*. Camberwell, Victoria: Australian Council for Educational Research.

Tytler, R. (2009). School innovation in science: Improving science teaching and learning in Australian schools. *International journal of science education, 31*(13), 1777–1809.

Urban-Woldron, H. (2014). Preparing prospective elementary teachers to teach science: A challenge for teacher education. *Open Online Journal for Research and Education, 11*(2).

Velthuis, C., Fisser, P., & Pieters, J. (2014). Teacher training and pre-service primary teachers' self-efficacy for science teaching. *Journal of Science Teacher Education, 25*(4), 445–464.

Vygotsky, L. (1977). The development of higher psychological functions. *Soviet Psychology, 16*(1), 60–73.

Wagler, R. (2011). Out in the field: Assessing the impact of vicarious experiences on preservice elementary science teaching efficacy. *Electronic Journal of Science Education, 15*(2), 1–28.

Watters, J. J. (2007, October 29–31). Problem based learning in pre-service elementary science teacher education: Hostile territory. *Proceedings of the PBL Conference on Problem-Based Learning in Undergraduate and Professional Education*. Birmingham, Alabama (pp. 1–10). Retrieved October 15, 2008, from http://www.eprints.qut.edu.au.

Watters, J. J., & Ginns, I. S. (1999). *Setting the environment for life-long learning: Collaborative and authentic learning practices in primary science teacher education*. Paper presented at the International Research Network "PACT" Conference, Chinese University of Hong Kong.

Watters, J., & Ginns, I. (2000). Developing motivation to teach elementary science: Effect of collaborative and authentic learning practises in preservice education. *Journal of Science Teacher Education, 11*(4), 301–321.

Wenner, G. (1995). Science knowledge and efficacy beliefs among preservice elementary teachers: A follow-up study. *Journal of Science Education and Technology, 4*(4), 307–315.

Wenner, G. (2001). Science and mathematics efficacy beliefs held by practicing and prospective teachers: A 5-year perspective. *Journal of Science Education and Technology, 10*(2), 181–187.

Wilson, S. (2012). Drivers and blockers: Embedding education for sustainability (EFS) in primary teacher education. *Australian Journal of Environmental Education, 28*(01), 42–56.

Wingfield, M. E., & Ramsey, J. (1999). *Improving science teaching self-efficacy of elementary pre-service teachers*. Paper presented at the Association for the Education of Teachers in Science Annual Meeting in Austin, TX.

Yang, E., Anderson, K. L., & Burke, B. (2014). The impact of service-learning on teacher candidates' self-efficacy in teaching STEM content to diverse learners. *International Journal of Research on Service-Learning in Teacher Education, 2*, 1–46.

Yılmaz, H., & Çavaş, P. (2008). The effect of the teaching practice on pre-service elementary teachers' science teaching efficacy and classroom management beliefs. *Eurasia Journal of Mathematics, Science and Technology Education, 4*(1), 45–54.

Yilmaz-Tuzun, O., & Topcu, M. S. (2008). Relationships among preservice science teachers' epistemological beliefs, epistemological world views, and self-efficacy beliefs. *International Journal of Science Education, 30*(1), 65–85.

Chapter 3
A Review of the Science Teaching Efficacy Belief Instrument A: In-service Teachers

Abstract At a time where scientifically literate citizens are needed to make informed decisions about global issues, the importance of science education cannot be overstated. Due to the wide array of stakeholders, perspectives and goals it can be challenging to find consistent trends across the literature. The Science Teaching Efficacy Belief Instrument A (STEBI-A) has proven to be a valid and reliable measure of teachers' science teaching efficacy beliefs for over 25 years. The purpose of this chapter is to consolidate the body of STEBI-A literature through a structured review of the methods and findings in this area of science education research. A total of 107 articles and dissertations were deemed to have used the STEBI-A instrument and were subsequently included in the analyses. The findings showed that the instrument has been employed in varied research designs across 15 different national contexts to provide valuable insights into the nature, growth and cross-variable relationships of teachers' science teaching efficacy beliefs. Analysis of interventions that were assessed via the STEBI-A showed professionally relevant, resource rich science programs could enhance the personal science teaching efficacy beliefs of in-service teachers across a multitude of contexts. Teachers often expressed cynical views as their personal science teaching efficacy scores were higher than their general science teaching outcome expectancies. Improving teachers' science teaching outcome expectancies should be an aim for future research. More implications are discussed within the chapter.

The Science Teaching Efficacy Belief Instrument—A

The Science Teaching Efficacy Belief Instrument A (STEBI-A) was developed to measure the personal and general aspects that comprise an in-service teacher's science teaching efficacy beliefs (Riggs and Enochs 1990). The constructs that comprise the two subscales of the instrument are Personal Science Teaching Efficacy beliefs (PSTE) and Science Teaching Outcome Expectancies (STOE). The PSTE subscale measures respondents' beliefs about their own capacity to deliver science teaching experiences that assist students to develop predetermined science skills

© The Author(s) 2017
J. Deehan, *The Science Teaching Efficacy Belief Instruments (STEBI A and B)*,
SpringerBriefs in Education, DOI 10.1007/978-3-319-42465-1_3

and knowledge. An example of an item on the PSTE subscale is "I find it difficult to explain to students why experiments work". The STOE subscale measures respondents' beliefs about the capacity of science teaching to overcome external factors to aid students' science learning in a general sense. An example of an item on the STOE subscale is "Students' achievement in science is directly related to their teacher's effectiveness in science teaching". The original STEBI-A is a 30-item Likert questionnaire wherein respondents rate their level of agreement with set statements on a 5-point scale ranging from 'strongly agree' to 'strongly disagree'. Over the course of its 25-year existence the STEBI-A instrument has been used to make meaningful contributions to a broad array of different research contexts.

The Research Contexts of the STEBI-A Instrument

The researcher acknowledges that technological limitations may mean that the research collected within this review may not be an exhaustive representation of the STEBI-A usage. Despite claims relating to the outdated nature of the science teaching efficacy beliefs instruments (Mulholland et al. 2004), use of the STEBI-A instrument has been rising dramatically. The STEBI-A instrument was adopted at a modest rate during its first decade of research use (1990–1999), with an average of 1 research item utilising the instrument per year. During the following decade (2000–2009), the use of the STEBI-A more than quadrupled to a rate of 4.4 research items per year. Over the past four years (2010–2014), the uptake of the STEBI-A has risen significantly to 12.75 research items per year. In fact, the number of research items employing the STEBI-A in the past 4 years (51) has already exceeded the entire output of the previous decade (44).

The overwhelming majority of the STEBI-A literature is situated within the American context. However, in recent years other nations have begun contributing to the existing body of STEBI-A literature. Figure 3.1 outlines the STEBI-A research output by nation. The United States accounts for 72 % of the STEBI-A research, with a total over seven times higher than the next most productive nation, Australia. In a similar trend to the STEBI-B instrument, the STEBI-A instrument has been used consistently within Turkish contexts over the past decade. A promising finding was the emerging trend of STEBI-A usage within differing cultural, economic and educational contexts. The STEBI-A instrument has been introduced to following national contexts: China (Sang et al. 2012), Denmark (Andersen et al. 2004), Ecuador (Lucero et al. 2013), India (Desouza et al. 2004), Iran (Fathi-Azar 2002), Israel (Eshach 2003), Netherlands (Velthuis 2014), Taiwan (Liu et al. 2008) and the United Arab Emirates (McKinnon et al. 2014). Such widespread use of the STEBI-A instrument may be related to global recognition of the importance of science education. There may also be opportunities in the future for comparative STEBI-A research across multiple contexts (e.g. Batiza et al. 2013). This will be unpacked further in the discussion section.

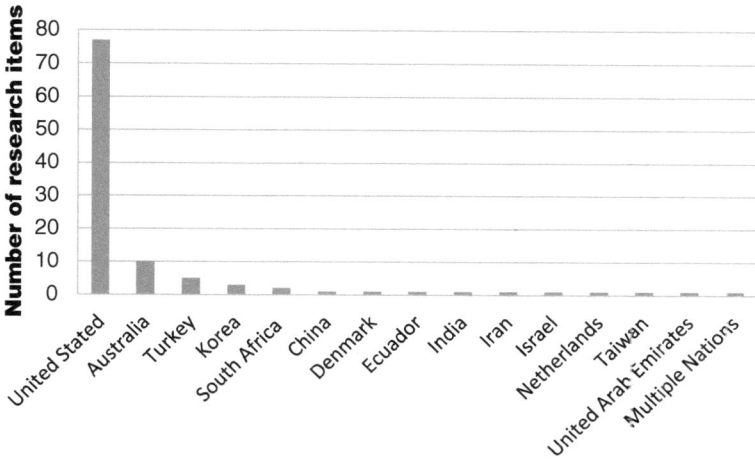

Fig. 3.1 Summary of the STEBI-A use of different nations

Purpose of This Chapter

There are three aims for this chapter. The first aim is to organise and describe the research employing the STEBI-A instrument from a methodological perspective. The second aim is to provide an overview of how the STEBI-A subscales have been used within the literature. The final aim is to explore and rank the innovative practices used within science interventions through the use of STEBI-A effect sizes as a comparison point. All of these aims will be achieved within the organisation framework presented in the previous STEBI-B review chapter.

Method

This chapter reports on a review of multiple research items that utilise the STEBI-A instrument. Such an approach allows for a clear understanding of how the research in science education has developed over the past 25 years. A variety of literature search techniques were used to acquire research items that have employed the STEBI-A instrument since its development in 1990. The acquired research items were coded for comparison in the following areas: context, research design, intervention, delivery, participants, use of subscales (PSTE and STOE) and effect sizes (Cohen's D). The following subsections will describe both the literature search procedures and the subsequent coding and analytic procedures used with relevant research items.

Inclusion Criteria

A Broad inclusion criterion (Suter 2006) was used for the literature search. This allows for research beyond the author's conceptualisations to be considered. The criterion was:

- The STEBI-A instrument is used to inform the research in a meaningful way. This accounts for the diverse contexts within which the STEBI-A can be employed.

Literature Collection Procedures

The initial search for literature occurred in 4 phases:

1. **Seminal author search**—The article discussing the development of the STEBI-A instrument by the seminal authors (Kervin et al. 2006) was found using 'Google Scholar'.
   Riggs and Enochs (1990). Toward the development of an elementary teacher's science teaching efficacy belief instrument. *Science Education, 74*(6), 625–637. Articles that referenced the seminal STEBI-A authors were tracked through a backward mapping approach (Green et al. 2006). The method yielded 116 relevant articles.
2. **ERIC Search**—The Education Resource Information Centre database was specifically searched with the use of the terms 'STEBI-A' and 'Science Teaching Efficacy Belief Instrument A'. An additional 5 articles were found that appeared to fulfil the inclusion criteria for the review.
3. **Primo Search**—The Charles Sturt University Primo search website was used to search for relevant literature. Primo affords access to an assortment of journals, newspaper articles, books, Ebooks and CSU research output. 'STEBI-A' and 'Science Teaching Efficacy Belief Instrument A' were utilised for this search. The Primo Search uncovered 9 research items that appeared relevant to the review.
4. **EBSCO Host Search**—The EBSCO Host search allowed the research to gain access to over 200 education based academic journals. The key search terms of 'STEBI-A' and 'Science Teaching Efficacy Belief Instrument A' returned 2 relevant articles that incorporated the STEBI-A instrument.
5. **Minor Journal Searches**—An array of minor research data bases were searched via the key terms of 'STEBI-A' and 'Science Teaching Efficacy Belief Instrument A'. The minor data bases included; Cambridge Journals, CBCA Database, Emerald, Expanded Academic ASAP, Infotrac, Factiva, Informit, Web of Knowledge, JSTOR, Oxford Journals, ProQuest, SAGE Journals Online, ScienceDirect (Elsevier SD), Scopus, Springerlink, Taylor and Francis Online, and Wiley Online Library. Ten additional relevant items were found amongst these journals.
   A total of 142 relevant research items were collected during the literature collection phase. When repeated and inappropriate articles were removed, the number of articles included in the analysis was reduced to 124. An additional 17 articles were removed as the STEBI-A had been incorrectly administered to pre-service teachers. A total of 107 research items were included in the review.

Coding and Analysis

The following subsections outline the coding and analysis procedures used for the STEBI-A papers. Coding and analytic procedures will be outlined for research approaches, interventions, and the use of Effect sizes.

Research Approaches

The research items were coded based on the methodologies within which the STEBI-A was employed. Qualitative use of the STEBI-A was coded as a zero, due to the difficulty in relating findings to alternate contexts. Quantitative or mixed methods research designs were coded based on the number of STEBI-A administrations. It should be noted that the codes do not represent judgements on the quality of the different research designs. Table 3.1 describes the research approaches and codes for the targeted research items.

To supplement the research design analyses, the research items were coded in additional ways. Firstly, the subscale differentiation (PSTE and STOE) was coded on a 3-point scale. A score of zero indicated that the subscale was not present, a score of 1 indicated that the subscale was merged, and a score of 2 indicated that the subscale was present. Secondly, the descriptive statistics of the qualifying studies were coded in terms of overall quality, within an underlying focus on the calculation of Cohen's D effect sizes for intra-study comparisons. A score of zero indicated that the descriptive statistics could not be found, a score of 1 indicated that the descriptive statistics were present but incomplete and a score of 2 indicated that the descriptive statistics were complete.

Intervention Coding

The articles that reported on a science intervention were coded based on the pedagogical innovations included in the educational design. A framework was initially developed from Lawrance and Palmers' (2003) description of innovative practices within tertiary science programs. The framework was modified to suit the in-service teaching context via a thorough manual analysis of the STEBI-A literature. Many of these innovations have overlapping elements but enough differentiation exists in the literature for separate definitions. The framework provides a useful, but not infallible, list of innovative practices. Other researchers should continue to make additions to and refine this list. Table 3.2 explains the selected innovative practices. A total of 15 approaches comprise this 'innovative practices' framework. The researcher acknowledges that this is not an exhaustive list of potential pedagogical approaches.

Table 3.1 Research methods codes

Code	Research type	Number of STEBI-A uses	Description
4	Experimental design	>2	Experimental designs allow for cause and effect statements to be made rather than correlational observations. This research design is comprised of two groups. The experimental group is exposed to a formal treatment. A control does not receive the treatment. Where possible extraneous variables are controlled through randomised group assignment. However, in educational research it is often ethically impossible to randomly assign participants to either group
3	Longitudinal quasi experimental pre/post design	>2	Research where a pre- and post-test STEBI-A implementation is supplemented by delayed testing to determine the longevity of any efficacious changes in the absence of formal science treatment
2.5	Quasi experimental pre/post—with multiple cohorts	>2	Research where multiple cohorts of participants respond to pre- and post-test versions of the STEBI-A, as they undertake a specified intervention
2	Quasi experimental pre/post	2	Research where a cohort of participants provide STEBI-A data, both before and after, undertaking a specified intervention
1.5	Equivalent groups	2	Research where pre and post intervention STEBI-A data are collected from separate groups and compared as equivalent data (Suter 2006)
1	Cross sectional	1	Research where the STEBI-A was administered to a single group on one occasion to make comparisons with other variables
0	Qualitative/alternate research approaches	Variable	There are two primary types of research that fall into this category Firstly, research where the STEBI-A instrument was not used to provide statistical data, as originally intended by the seminal authors (i.e. qualitative research) (Riggs and Enochs 1990). Secondly, research where the STEBI-A instrument was outlined in the methodology, but the subsequent STEBI-A data was not presented

Table 3.2 Overview of Innovative practices used within the analyses

Innovative practice	Description
Constructivism	Learning that occurs when an individual constructs their knowledge through active participation (i.e. discussion) within a phenomenon or situation (Slavin1991; Vygotsky 1977)
Problem-based learning	Problem-based learning is a deep learning strategy that helps students to develop transferrable skills, which can be used in novel situations (Schmidt et al. 2006). Problem-based learning uses real-world problems as a starting point for the acquisition and integration of new knowledge into existing schemas (Azer 2001; Kahn and O'Rourke 2005)
Integration with other Key Learning Areas (KLAs)	An approach to teaching where two disciplines, that are considered fundamentally separate, are integrated to create deep learning outcomes. For example, allowing students to collect and graph data is an example of a deep integration between mathematics and science. For the purposes of this analysis, this 'innovative practice' was split into two sub-categories: 'Deep links to mathematics (STEM)' and 'Integration with other KLAs
Mentoring	Mentoring is an emerging practice where teachers are paired with more experienced expert teachers or outside experts in order to focus on a particular discipline. The teachers observe the experienced teachers and receive feedback on their own teaching practice
Curriculum development/ real world relevance	This term broadly encompasses teaching pedagogies and learning opportunities that accurately reflect the responsibilities and actions of the teaching profession. Within the context of this review paper this would include approaches such as allowing teachers to develop units of work and other resources that can be used in their own teaching practice
Inquiry learning	Inquiry learning allows participants to develop transferable skills and knowledge to seek the information needed in order to achieve a task (Duran et al. 2009; Edelson et al. 1999). Open inquiry occurs when the participants have complete control over inquiry processes. Guided inquiry occurs when some instruction and support is provided
School partnership	This occurs when the intervention is designed with imbedded opportunities to teach science to students of the intended year levels. The research item may describe an ongoing relationship between the group providing the science intervention and school groups
Tertiary partnership	This occurs when there is an explicitly stated relationship between the intervention providers and a pre-service teaching institution. Such a relationship allows for tertiary benefits (science subjects, on-campus resources, and access to relevant experts) to be received by participating in-service teachers
Cooperative learning	Cooperative learning occurs when teachers work together in separate, complimentary roles to complete a task that would otherwise be impossible to complete individually
ICT instruction/ incorporation	The teachers explicitly learn about the use of ICT in a way that is relevant to classroom teaching practice. For example, the use of Interactive Whiteboard Software for creating learning aids

(continued)

Table 3.2 (continued)

Innovative practice	Description
Student centred investigation	These are investigations where the teachers assume the locus of control within the confines of the subject. Ideally, teachers should have control over all stages of the investigation, with the instructor acting in a facilitative role
Authentic tasks/real word experiences	Authentic tasks are those that are clearly related to the profession/career that the students are studying to enter. Examples may include; developing units of work, practical experiences, researching student misconceptions. Real world experiences refer to learning experiences where the participant engages in scientific practices which either model true scientific practice and/or present science with real world applications
Nature of science	The understanding that scientific knowledge is fluid and always subject to reasonable debate. Instruction in this area may orient the learner to the variety of scientific approaches beyond an experimental research design
Misconceptions	A misconceptions based approach is a practice where the misconceptions of the students/teachers are identified and revealed to them. Such misconceptions inform the develop of personalised curricula. This approach can be used to aid a teacher science intervention whilst serving as a model for in class science teaching practice

The interventions were coded dichotomously as either including (1) or not including (0) each innovative practice. The judgement was based upon the author's thorough reading of the intervention descriptions, which were supplemented by the use of a search function to assess the inclusion of key terms. An innovative practice did not have to be explicitly explained within an intervention to be classified as 'included', rather the practice had to be evident within the description based on the informed reading of the researcher. The quality and depth of innovative practices were not differentiated in this coding scheme. The author acknowledges that the coding of each intervention may not be complete as practices may have been missed or misinterpreted. The interventions were also coded for both the delivery and length of the programs. The delivery was coded at 3 levels: online (1), face-to-face (2) and mixed (3). The program lengths were coded as; single workshop (2), multiple workshops (2), and multiple workshops with in-school support (3).

Using Effect Sizes to Evaluate the Interventions

STEBI-A research that administered the instrument to the participants at least twice (coded ≥ 1.5 for research design) over the course of a specified intervention were analysed for Cohen's D effect sizes. This allowed the research to determine which interventions, and the corresponding pedagogical approaches, correlated with the largest effect size increases for the PSTE and STOE scores of participating in-service teachers. For research that reported on two or more cohorts, a mean was calculated in one of two ways. Firstly, when no outlier was present a mean effect size was calculated. Secondly, when an outlier (> or <0.5 above other cohorts) was present in research reporting on 3 or more cohorts, the median score of all cohorts was used to represent the research.

Structural Framework for the Analysis of the STEBI-A Literature

The structural framework was introduced in the first chapter as a means to succinctly describe and analyse research employing the Science Teaching Efficacy Belief Instrument B (STEBI-B) (Enochs and Riggs 1990). As the STEBI-B and STEBI-A instruments measure the same constructs with minor contextual modifications, the structural framework can be applied to both instruments. Figure 3.2 shows the framework for the STEBI-A analysis. The blue sections show the research designs and the red sections show the innovations/effect size analyses. The research designs outlined in the inverted blue triangle are arranged in approximate order of the number of STEBI-A uses. The red triangle represents evaluative research designs that assessed a science intervention with at least two administrations of the STEBI-A instrument. The red arrows show the information that was extracted from these evaluative research articles. The left circle of the Venn diagram shows

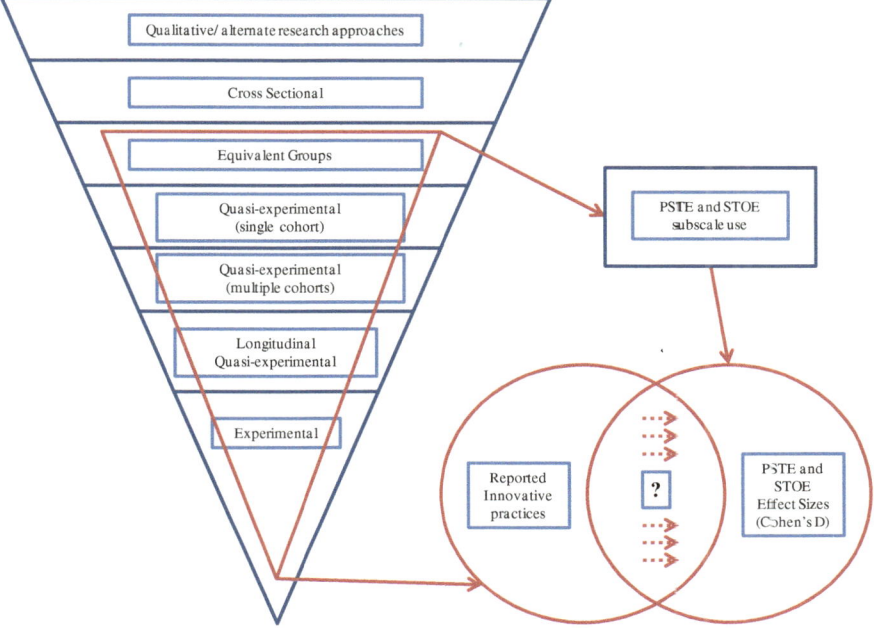

Fig. 3.2 STEBI-A structural review framework

the innovative practices reported in the relevant literature. The right circle shows the STEBI-A effect sizes reported within the analysed literature. It should be noted that the effect sizes are moderated by researchers' usage of the PSTE and STOE subscales. The intersection of the Venn diagram acknowledges and represents the unknowable relationship between the interventions and the report effect sizes.

Findings

The findings will be presented as separate research design and innovative practices sections. Firstly, the STEBI-A research design section will provide a holistic overview of how the instrument has been employed within the literature. Secondly, the focus will tighten to focus on research that describes science interventions and/or evaluates those interventions with STEB effect sizes.

STEBI-A Research Designs

The following subsections will describe the different research designs that have used the STEBI-A instrument. The research designs have been evaluated based on the both the use of the STEBI-A instrument and the number of administrations. A funnel has been created as articles and dissertations are removed at the end of each section. All 107 STEBI-A research projects were considered in this analysis.

Qualitative Research and Alternate STEBI-A Uses (0)

Many researchers have chosen to use the STEBI-A in qualitative research designs. A steady flow of qualitative research has been published with the intent to bridge the gap between teachers' pre-service and in-service science teaching efficacy beliefs. Ginns and Watters (1996) followed the early career experiences of two case study participants. The results showed that the school environment (experienced teachers and administrators) played a crucial role in developing the science teaching efficacy beliefs of novice teachers. Saka (2007) also found that the professional school contexts played an important role in shaping the science teaching practices and efficacy of two early career teachers. Contradictions between university and professional school contexts appear to place considerable stress on early career teachers. In fact, one participant in Saka's (2007) research changed his personal goal for science reform due to such contradictions. Ginns and Watters (1999) used the STEBI-A to frame interview and video data collected from four case study teachers. It was found that the four novice teachers all drew heavily from their tertiary experiences to develop constructivist, student centred science curricula. Qualitative STEBI-A research often serves to illuminate issues that arise during the transition from pre-service to in-service status. Yet, deeper quantitative and mixed method research is needed to complement these inductive trends. Larger sample sizes are needed to ensure the generalisability of findings in the future.

Researcher choices in terms of STEBI-A usage and reporting have lead to a number of research items being classified as alternate. While this does limit the relevance of these studies in the context of this review, it does not diminish wider contributions. Some researchers chose not to report STEBI-A results despite identifying the instrument as a tool for data collection (Duran et al. 2010; Shroyer et al. 1996). The choice to omit STEBI-A data is not always problematic. For example, Urquhart and Bober (2006) produced a paper with the primary aim of describing a Physics based professional development course rather than evaluating it. The STEBI-A was described as an evaluative tool within the course design. In another circumstance, the STEBI-A was modified to an extent that greatly reduced the capacity of the research to be compared with other STEBI-A research. Specifically, Taylor (2005) created a combined Environmental and General Science Teacher Efficacy instrument by combining items from the STEBI-A with other instruments. Ultimately, researchers' data collection and reporting choices will always be informed by factors of context and publication requirements. While such choices may appear to limit the research in the context of this review, it does not diminish the overarching contribution of these papers to the literature base.

Science interventions and other forms of professional development are the subjects of some papers coded at the qualitative/alternate level. Waters and Ginns (1997) explored how a single teacher responded to a 'Simply Science' program where 25-minute science television programs were broadcast to students and supplemented by follow-up science lessons. While the experienced teacher showed no change in her science teaching efficacy beliefs, she did show improved Pedagogical Content Knowledge (PCK) and better attitudes to constructivist teaching approaches. The importance of PCK has been reaffirmed in later research. Nafziger (2008) found that the self-efficacy of three teachers participating in a standards-based, inquiry science program only improved if their developing science content knowledge could be connected directly to their teaching practice. Saka et al. (2009) found that science teaching efficacy was not a key blocking factor in 3 teachers' enactment of inquiry teaching practice. Instead pedagogical discontentment, perceptions of students' ability levels and external contextual pressures were identified as key factors that needed to be addressed to ensure the enactment of inquiry teaching. Ogbomo (2010) reported on a museum based, science professional development program that made clear contextual links for teachers in terms of resources and curricular relevance. These inclusions appeared to produce positive results as the six teachers reported improve science teaching efficacy and science content knowledge. The main point of contention was the absence of in-school follow-up research. The average number of participants in science intervention research at this level is nine. This means that the findings cannot be generalized to form more definitive conclusions. Nevertheless, the rich findings of the research at this level have informed and will continue to inform the direction of future research. A total of 14 pieces of research used the STEBI-A instrument in either qualitative or alternate research designs. The remaining 93 studies were considered in the next step of the research designs analysis.

Cross Sectional Research Designs (1)

Much of the cross sectional STEBI-A research explores the relationship between science teaching efficacy and classroom practices. Burton (1996) found that statistically significant positive correlations existed between the science teaching efficacy beliefs and reported constructivist science teaching practices of a sample of 285 American elementary teachers. Lardy (2011) found a similar positive correlation between science teaching efficacy and reported constructivist teaching beliefs. Classroom observations showed that these beliefs were not evident in science teaching practice. Secondary school science teachers may have similar preferences for transmissive, teacher centred approaches. In a sample of 86 secondary science teachers, those with high and low science teaching efficacy beliefs displayed similar preferences for traditional methods of instruction (Hodgin 2014). Lucero et al. (2013) suggest that teachers with higher personal science teaching efficacy may be more likely to give students autonomy in their inquiry learning. Highly structured and supported professional development opportunities can improve teachers' science teaching efficacy and the frequency of their science teaching practice. Albion and Spence (2013) found that teachers' who used Primary Connections curriculum materials reported both higher science teaching efficacy scores and high science teaching frequency than teachers' who did not have access to the same materials and support.

 Cross sectional STEBI-A research has been used to assess the relationships between science teachers and other stakeholders. In recognition of the core goals of both elementary and secondary science education, some STEBI-A research connects teacher science efficacy with student characteristics and learning outcomes. A survey of 225 teachers showed that there was no significant relationship between teachers' STOE scores and their students' economic level or ethnicity. Curiously, teachers who classified their students as coming from a middle/upper income background had higher PSTE scores than their counterparts who classified their students as coming from low income backgrounds. Science teaching efficacy may be partially related to student outcomes. Angle and Moseley (2009) assessed the science teaching efficacy beliefs of secondary biology teachers whose students' outcomes both fell below the proficient level and reached the proficient level. The teachers with students performing at proficient levels reported higher STOE scores. There was no difference in PSTE based on student achievement level. Other research suggests school leadership is related to teachers' PSTE scores. Clark (2009) found that teachers with higher PSTE scores described their principals as engaging in positive leadership practices such as: staff goal discussion, science curriculum reviews, recognition of student progress and the encouragement of innovative teaching approaches.

Education variables such as content knowledge and tertiary experiences have underpinned some of the cross sectional STEBI-A research. Desouza et al. (2004) administered the STEBI-A instrument and a demographic survey to over 300 Indian middle school teachers. There was a strong positive correlation between teachers' level of education and their STEB scores. A curious finding was that the correlation between the PSTE and STOE subscales weakened as the level of teacher education attainment increased. In fact, there was no significant correlation between PSTE and STOE for the teachers with Bachelors' Degrees but no preparatory practical science teaching experience. Lekhu (2013) found that there was no relationship between the science qualifications and science teaching efficacy beliefs for 99 South African secondary science teachers. It was noted that higher STOE scores were associated with higher professional teaching qualifications. In a sample of 113 primary teachers, Saint (2013) conducted regression analyses to find that pre-service science education experiences account for 34 % of the variance in teachers' science teaching efficacy belief scores. In-service professional development accounted for 39 % of the variance in STEBI-A scores. Both pre-service and in-service training showed significant predictive relationships with science teaching efficacy beliefs. After the 38 cross sectional studies were eliminated, the pool of research papers decreased to 55.

Research with Equivalent Groups (1.5)

None of the analysed articles employed an equivalent groups design.

Quasi Experimental Designs with a Single Cohort (2)

A large proportion of the quasi experimental STEBI-A research evaluates the effectiveness of professional development programs. Professional development practices such as: peer teaching (Finson et al. 1992); field trip experiences (Kean and Enochs 2001); practical teaching experiences (Naizer et al. 2003); the 5Es framework (Shea et al. 2013) and problem-based learning (Ertmer et al. 2014) have all been evaluated through pre- and post-test administrations of the STEBI-A instrument. To avoid redundancy this section will discuss the quasi-experimental, single cohort research with smaller numbers of participants. Holbert et al. (2011) aimed to foster relationships between schools and universities by offering research experiences for 14 experienced teachers. After completing the program the participants reported increased interest in science and mathematics research. There was evidence of a rift between schools and universities as the respondents displayed no science teaching efficacy growth and were critical of the performance of their university instructors. Thomas et al. (2013) described a Geology content program delivered by tertiary experts. The seven participants showed large effect size increases in their PSTE scores, but their STOE growth was negligible. Rather than addressing science content knowledge deficits, Pinnell et al. (2013) described a STEM education framework that engaged 10 in-service teachers in curriculum development and research experiences. The STEBI-A results were the inverse of those reported by Thomas et al. (2013) as the STOE showed significant change whilst the PSTE remained stagnant. The sample size of 6 teachers on the post-test is severely limiting in this research as statistical assumptions cannot be met.

The research at this level shows an emerging trend of long-term professional development delivery with increased opportunities for participants to connect with instructors. Haney et al. (2007) reported on the teacher outcomes from participation in a 2-year professional development program wherein participants developed integrated science programs in a problem-based learning environment. The results aligned with broader literature trends as the 18 participants displayed increased PSTE scores, improved attitudes towards non-traditional science teaching practices and small, but significant, declines in their STOE scores. Kuchey et al. (2009) describe a similarly structured 2-year professional development program where the 18 teachers participated in monthly workshops. The program was content driven, used constructivist approaches and integrated mathematics with science. The teachers displayed strong increases in their science teaching efficacy beliefs, but the merging of the subscales prevents meaningful comparisons to other STEBI-A research. The long-term delivery of the professional development allowed the schools to complete site based science and mathematics improvement programs to extend beyond the duration of the program itself. Other programs have been supplemented with online learning opportunities to foster comparably deep links for teacher participants. Gosselin et al. (2010) described and evaluated the online 'Laboratory Earth' program. The 51 participating teachers had access to online modules, support materials, instructor support and discussion forums to allow for effective synchronous and asynchronous learning. The results showed both science teaching efficacy and content knowledge increases for the participating teachers. Other programs have included online components to supplement science professional development programs. Online meeting sessions have been used to facilitate curriculum development opportunities for practicing teachers (Holbert et al. 2011). Online assessment and feedback have also been used to recognise the prior knowledge of in-service teachers (Rudman and Webb 2009). It appears as though the STEBI-A research is only just beginning to explore the potential of long-term professional development models supplemented by online delivery. There were 22 studies that used the STEBI-A instrument to record pre- and post-test data from a single cohort of participants. There were 33 research pieces that were carried over to the next level of analysis.

Quasi Experimental Designs with Multiple Cohorts (2.5)

Research where STEBI-A data has been collected from multiple cohorts affords excellent opportunities for "between group" comparisons. Some research reveals consistent STEB growth in participants across numerous iterations of science programs. Roberts et al. (2001) found that an inquiry-based science professional development program produced large PSTE effect size changes for the 330 teachers who participated between 1992 through to 1999. Of note was that there were no statistically significant differences between the year groups despite changes to the length of the program. Shin et al. (2010) described remarkably uniform small-to-moderate science teaching efficacy gains in the 75 teachers who participated in a problem-based learning professional development program. In fact, across the four years of data collection (2006–2009) antecedent variables such as teaching experience and previous professional development had no significant impact on participants' science teaching efficacy scores. In other cases, research at this level (2.5) highlights inconsistencies in the STEB results across different iterations of science professional development programs. For example, two cohorts involved in a teacher centred systemic reform model showed considerable disparities in their PSTE effect sizes, with the first group showing no change but the second group showing large effect size growth (Saka et al. 2009). Similar inconsistency between group scores has been reported on the STOE subscale (Lockman 2006). Nevertheless, multiple cohort research designs strengthen arguments for covariant relationships between teachers' science teaching efficacy growth and participation in science professional development. Such approaches make rich contributions to the STEBI-A literature by helping to rectify some of the issues associated with the absence of true experimental designs.

Some research at this level presented merged STEBI-A subscales. The choice to merge the subscales prevents between studies comparisons but the research provides some notable standalone information. Nadelson et al. (2013) used a merged science teaching efficacy belief score, amongst other measures, to evaluate the impact of an inquiry-based STEM professional development program. They reported on two separate offerings of the program over two years, with the first cohort providing data in year one and the second cohort providing data in year two. The first cohort showed a large overall STEB change (Cohen's $D = 0.923$). The second cohort displayed only moderate STEB growth (Cohen's $D = 0.452$). Both groups reported improved attitudes towards STEM and greater science teaching confidence. Perhaps most importantly, many of the 68 teachers displayed a heightened inclination to incorporate STEM concepts in their classroom curricula. Ellins et al. (2013) describe a 5-year teacher professional development program entitled 'The Texas Earth and Space Science (TXESS) Revolution'. Participating secondary science teachers were provided with mentoring, curriculum support and workshops to develop their geoscience pedagogical content knowledge. The participants reported greater geoscience content knowledge and increased confidence for the classroom use of technology. The STEBI-A use was problematic as the subscales were merged and the broad mean scores were not differentiated for the 3 iterations of the program. Nonetheless, the teachers' science teaching efficacy beliefs did not display significant growth. There were 11 research items that reported on pre- and post-test data from multiple cohorts. 22 articles progressed to the next level of research design analysis.

Longitudinal Quasi-Experimental Research Designs (3)

Longitudinal STEBI-A research has been used to elucidate the, often overlooked, experiences of early career teachers as they transition from pre-service to in-service status. Wingfield et al. (2000) used a longitudinal design to determine if pre-service science teaching efficacy beliefs were maintained through the first year of in-service teaching. The results indicated that the PSTE and STOE growth which occurred within the 31 participants of a site-based tertiary science program remained durable after their first year of in-service teaching. Similarly, Palmer (2011) found that the strong PSTE effect size gains made by pre-service teachers, who participated in a science intervention, incorporating cognitive mastery, enactive mastery, modelling and verbal persuasion, remained durable for two years after the conclusion of the intervention. It is important to recognise that school variables can potentially confound the durable efficacious impacts of tertiary science interventions. Andersen et al. (2004) collected longitudinal STEBI-A data from 66 Danish elementary teachers at the beginning, middle and end of their first year of teaching in order to determine how their science teaching efficacy beliefs were influenced by teaching environment variables. Analyses showed that there was a relationship with participants' PSTE scores and their beliefs about their school science contexts. This relationship is of particularly interest as the first year teachers showed small decreases in their PSTE scores (Cohen's $D = -0.36$). This may be construed as evidence for a negative science culture existing at the elementary school level.

Longitudinal STEBI-A research designs have been used to evaluate established ongoing science professional development programs. A group of 12 teacher leaders received a combination of content instruction, problem-based Science instruction and leadership instruction over a 3-year period (Mentzer et al. 2014). At the completion of the study, the teacher leaders reported statistically significant increases in their science content knowledge and science teaching outcome expectancy scores. The stagnation of the PSTE scores defies broader trends within the STEBI-A literature. Ulmer et al. (2013) found that an integrated Agriculture professional development program, with a focus on curriculum development, seemed to produce results that aligned more closely with the common tendencies in the STEBI-A literature base. The large PSTE effect size gains (Cohen's $D = 0.958$) remained durable after the program had ended. The STOE effect size increases were not durable as the final scores were not significantly different from the pre-test scores. In a comparable study, Sandholtz and Ringstaff (2014) collected data over a 3-year period to assess the impact of a research-based science content course on 39 K − 2 teachers. The authors reported strong, durable increases to both the PSTE and STOE scores of the participants. However, these findings could not be verified as the descriptive statistics were omitted from the article. There were 8 research articles remaining after the previous 14 were removed.

Experimental Research Designs (4)

The remaining 8 research items employed approximate experimental research designs. That is to say that both experimental and control groups were clearly identified. It should be noted that random group assignment was not always a feature of the research described at this level. McConnell et al (2008) employed an experimental design to determine if video-based reflective practices enhanced teachers' capacity to make evidence based decisions relating to their practice and improved their science teaching efficacy beliefs. The experimental group showed greater improvements in their science teaching efficacy beliefs and showed more scepticism toward memory-based reflection as they began to favour evidence-based reflection to inform pedagogical choices. In fact, while the experimental group showed large effect size increases (Cohen's $D = 0.98$) to their PSTE scores, the control group showed moderate declines on the same subscale (Cohen's $D = -0.63$). The generalisability of these STEB-A data are limited by the total sample size of 15 participants. This research could therefore not be considered in other sections of this review. Sang et al. (2012) investigated the modality of pro-fessional development delivery in a randomly grouped, experimental research project. The experimental group participated in a 10-week constructivist science program that required deep analysis of video-taped science lessons from novice and expert science teachers. The control group provided the same pre- and post-test data but did not undertake an intervention. The experimental groups showed moderate PSTE increases. Resentful demoralisation may have hindered the valid-ity of the research as the control group displayed statistically significant declines on both the PSTE and STOE subscales. The implication may be that alternate interventions in a modified experimental design may be more appropriate for STEBI-A research than a control group experiencing no intervention because of the expectations and perceptions of participants. Fishman et al. (2013) overcame such difficulties by delivering similar professional development experiences with the key differentiating variable of delivery method. A total of 49 second-ary teachers were assigned to either face–to–face or online delivery groups of an inquiry-based environment science professional development program. Regression analyses showed that there were no significant differences on science teaching efficacy between groups. Cohen's D effect sizes could not be calculated due to the absence of key descriptive statistics. The following section will analyse the use of the PSTE and STOE subscales within the STEBI-A literature.

The PSTE and STOE Subscales Within the STEBI-A Literature

There appears to be some inconsistency in the usage of the subscales within STEBI-A literature. The STEBI-A subscales were used explicitly in 80 of the analysed research items. The PSTE subscale was used on all such occasions. Merging subscales and STOE omission were seen within the STEBI-A literature. Despite the complementary and separate nature of the PSTE and STOE subscales (Riggs and Enochs 1990), the two were merged in 14 research projects. Several of these research items recognised the different subscales but choose to provide merged means (e.g. Buyuktaskapu 2010; Ogbomo 2010; Saka 2007). More commonly, paired T-tests have been used to explore changes in participants' responses to individual items over time (e.g. Duran et al. 2009; Kuchey et al. 2009; Sandholtz and Ringstaff 2011). Such approaches allow for analysis of the deeper variables that contribute to the STEB constructs but they hinder comparisons between research items from different contexts. In 20 % of the relevant articles the STOE subscale was simply omitted entirely. The choice to avoid the STOE subscale has become more prominent in the past decade as the 12 of the 16 articles/dissertations that fulfil this criterion have emerged after 2005 (e.g. Khourey-Bowers and Fenk 2009). The implications of the decline in STOE usage will be unpacked in the discussion. The relative difficulty in affecting STOE growth may be a factor in choices to avoid the subscale.

Within the STEBI-A literature teachers commonly show higher scores on the PSTE scale. Three quarters of the STEBI-A research makes clear distinctions between the reporting of PSTE and STOE results, thus allowing for clearer inter-research comparisons. In fact, there were 64 research items that accurately measured both the PSTE and STOE scores of the research participants. A total of 38 research items presented the data in a way that allowed for PSTE mean scores to be compared with STOE mean scores on at least one occasion (cross sectional). A clear majority (82 %) of these research items reported that the PSTE scores of participants were significantly higher than their STOE scores. This trend may be related to the higher number of latent variables, such as other teachers and school contexts, which can influence the STOE subscale. It is concerning that teachers' perceptions of their own capacity to deliver quality science teaching experiences are higher than their own expectations of science teaching outcomes in a general sense. These trends may hint at an unnerving cynicism within science teachers.

There were 24 examples of research where the growth rates of the subscales were comparable. This number is significantly less than the 65 articles that fulfilled these requirements in the STEBI-B review in Chap. 1. This may hint at the relative difficulty in obtaining consistent participant samples from populations of practicing teachers. Figure 3.3 shows the trends in STOE subscale growth for the STEBI-A instrument. The PSTE subscale displays more capacity for development, as 65 % of the analysed research items reported higher growth rates on the PSTE subscale. Higher STOE growth was reported in 25 % of the research projects,

Fig. 3.3 Trends in STEBI-A
subscale Growth

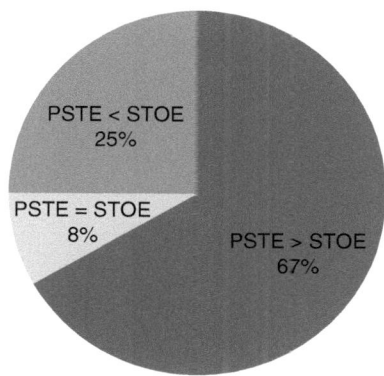

indicating that the STOE may be malleable for in-service teachers. This is up from 15 % reported for the STEBI-B instrument. The implications of this finding will be explored in the discussion section of this chapter.

There were two research items that reported comparable effect size gains on both subscales. McConnell et al. (2008) found that the use of video-recording as a tool for teacher reflection on science practice caused 1-sigma growth in both PSTE and STOE scores. Through the use of a retrospective post-pre-test, quasi-experimental design, Ulmer et al. (2013) found that participants in a problem-based agricultural professional development program displayed large gains in both PSTE and STOE scores. Curiously, a nine month follow up revealed that the PSTE gains were durables whilst the STOE gains proved unsustainable. Nevertheless, the core message remains mired in cynicism as in-service teachers, much like their pre-service counterparts, generally feel more confident in their personal science teaching ability than they do in the capacity of science teaching to improve students' knowledge and skills.

Innovative Practices Within the Science Interventions

There were 76 STEBI-A research publications that noted the inclusion of a science intervention in the body of the writing. Of these articles and dissertations, six did not include sufficient description to allow for intervention coding. The remaining 70 articles were coding dichotomously as either including or not including each of the 16 identified innovative practices. Tertiary partnerships and school partnerships were included to account for the in-service context. The raw numbers for each innovative practice are presented in Fig. 3.4. The average science program included 4.6 innovative practices, indicating that many innovative practices are being combined to create intricate educational designs.

The most frequently identified innovative practices were inquiry learning (64 %), cooperative learning (52 %) and curriculum development (38 %).

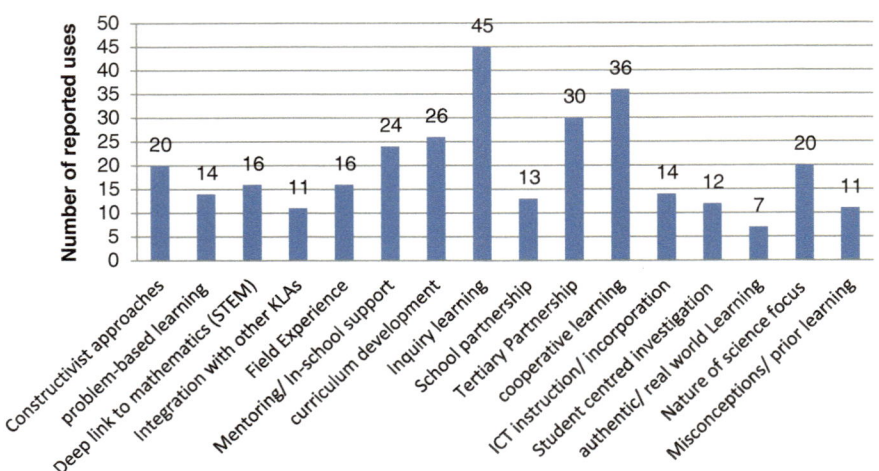

Fig. 3.4 Innovative practices included within the STEBI-A literature

This suggests that the professional development programs are tailored to suit the practical needs of practicing teachers. There were no innovative practices that were definitively absent from the coded science programs, indicating that science professional development may be more varied and reflexive than the science interventions offered to pre-service teachers at the tertiary level. Tertiary partnerships (43 %) and mentoring/in-school support (34 %) indicate that deeper, long-term programs are being delivered to account for the professional needs of in-service teachers. The researchers are recognising the potential benefits of participants' immediate access to appropriate elementary and secondary class groups. Such rich links to practical teaching opportunities are inherently more difficult to establish with pre-service teachers who often do not have access to a regular class.

When considering the innovative practices used within science professional development programs it is crucial to recognise the differences between pre-service and in-service teaching contexts. The aforementioned access to school contexts is an unquestionable advantage for the delivery of in-service science professional development, but there are difficulties that instructors need to overcome that are not as common in the tertiary context. The dispersal of in-service teachers amongst schools in combination with the professional requirements of the teaching profession makes it more difficult to contact and work with in-service teachers. To determine how professional development programs accounted for time restrictions, financial issues and geographic dispersal, all 69 interventions were coded based on both the type and length of program delivery. There were 66 programs that clearly stated the type of delivery. Each program was codes as either online, face-to-face or mixed. Face-to-face delivery was described in 53 articles (80 %); an understandable finding given the age of the STEBI-A instrument. Curiously, only two articles described an online delivery, one of which was published in 1996 (Shroyer et al. 1996). A new trend of mixing online and face-to-face learning opportunities appears to be emerging in the STEBI-A literature as 8 of the 11 programs that utilised this mixed approach were published after 2010. Online and mixed deliveries of professional development could play a role in addressing the disadvantages of those teachers in more isolated/rural school locations. 63 articles and dissertations were coded based on length of the professional development programs described (single contact, multiple contacts and multiple contacts with imbedded in-school follow ups). A promising finding was that only five programs described one contact session. The majority of professional development programs (54 %) incorporated multiple contact sessions for instructors and participants. 24 programs included follow-up school visits, reflections and/or clear links back to professional practice. Such trends may not be representative of science professional development as the coded programs have been taken from published research.

The PSTE and STOE Effect Sizes Produced by Science Interventions

Research reporting on data from less than 21 participants was removed from this section of the analysis. This ensures that statistical assumptions for the data set can be met as the pre- and post-test scores for each calculated effect size should have a relatively normal distribution of scores with minimal skewing. A total of 21 pieces of research were deemed suitable for PSTE and STOE subscale analysis. Figure 3.5 shows the distribution of effect sizes reported on both the PSTE and STOE subscales. The red lines differentiate insignificant, small, moderate, large and very large effect sizes. There is not a normal distribution of scores for either

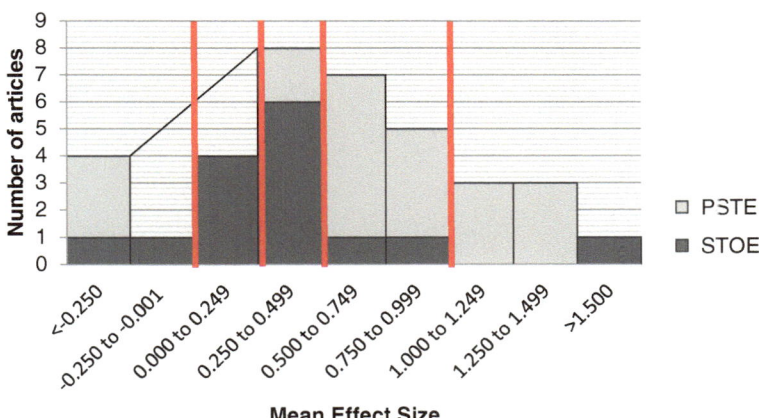

Fig. 3.5 PSTE and STOE distribution histogram

subscale. The high Kurtosis of both the PSTE (4.831) and STOE (9.737) subscales indicates that the spread of scores is notably separate from the mean point. When the extreme outliers are removed these Kurtosis scores drop to 1.927 and 2.256 respectively. The PSTE subscale has produced greater growth with many moderate to large effect sizes reported. The majority of STOE effect sizes are insignificant to small. This reflects the issues with the measurement and improvement of the STOE subscale that are rife within the STEBI-A and STEBI-B literature.

There is a significant difference between the mean PSTE and mean STOE effect sizes reported within the STEBI-A literature. The descriptive statistics for the subscale effect sizes are presented in Table 3.3. The mean PSTE effect size is greater than the mean STOE effect size. When the outliers are removed from the data set this gap widens as the mean PSTE effect size (0.77) is three times larger than the almost insignificant mean effect size (0.205) of the STOE subscale. The disparity may be larger as six studies that measured the growth of the PSTE subscale did not report on the STOE subscale scores. Simply put, there may be lower STOE effect sizes that are not presented in the STEBI-A literature. The subscale differences reflect the aforementioned disparity in the subscale mean scores where the STOE shows lower mean scores and an elevated propensity for stagnation. The discrepancy between the subscales closely resembles the trends within the STEBI-B literature.

The mean score and effect size differences between the PSTE and STOE subscales are reinforced by correlation analysis as a significant negative or positive correlation does not exist between the efficacy measures. Table 3.4 presents the output for the PSTE and STOE effect size correlation analysis. The correlation analysis suggests that there is almost no correlation between PSTE and STOE effect sizes (Pearson's $R = 0.121$). This result is notably different from the STEBI-B analysis, which showed a moderate-to-strong correlation between the subscale effect sizes (Pearson's $R = 0.628$). The small sample size of relevant effect size data for the STEBI-A analysis may be a factor in this result. The result itself may suggest that the PSTE and STOE constructs separate further when

Table 3.3 Descriptive statistics for PSTE and STOE effect sizes

	Mean	Std. deviation	N
MeanPSTE	0.572	0.757	21
MeanSTOE	0.398	0.817	15

Table 3.4 Correlation analysis for PSTE and STOE effect sizes

		MeanPSTE	MeanSTOE
MeanPSTE	Pearson correlation	1	0.121[*]
	Sig. (2-tailed)		0.667
	N	21	15
MeanSTOE	Pearson correlation	0.121[*]	1
	Sig. (2-tailed)	0.667	
	N	15	15

*Correlation is significant at the 0.05 level (2-tailed)

pre-service teachers graduate and enter the teaching profession. Certainly, experienced teachers have access to a wider variety of experiences and perspectives that can affect their general outcome expectancies. The limited sample of research prevents definitive statements being made here. The following paragraphs will rank the PSTE and STOE effect sizes produced for in-service teachers and broadly describe the pedagogical themes. Due to the low number of articles, the top 5 PSTE and STOE research projects will be presented and discussed. This prevents insignificant STOE effect sizes from being presented.

The innovative practices used by the top five science interventions based on PSTE effect sizes produced mirror the trends within the wider body of coded science programs. Inquiry learning (3), teaching experiences (3) and curriculum development (2) were the most common innovative practices amongst this group. Table 3.5 ranks the top research pieces on PSTE effect sizes produced. A curious finding was that none of the 5 research projects featured a tertiary partnership as an overt programmatic inclusion. This could be hinting at recognition of the differing needs of pre-service and in-service teachers. It may also be related to the practical issues associated with delivering professional development through a tertiary institution. The complexity of the research designs is lower amongst this group, with an average of 4 innovative practices compared to the mean of 4.6 amongst all coded science interventions. Perhaps the demands of in-service teaching call for clearer and deeper science interventions.

All of the top 5 PSTE interventions reported strong effect size increases within participants. Roberts et al. (2001) collected data over an 8-year period to assess how different delivery timeframes influenced the personal science teaching efficacy beliefs of over 300 teachers. The program employed a constructivist approach to enhance participants' understanding of inquiry pedagogies. They were also afforded practical teaching opportunities and mentoring sessions with experienced science teachers and scientists. A large PSTE effect size (Cohen's D = 1.3) was reported across the four groups. An interesting finding was that a four week delivery period produced the greatest PSTE gains, with longer periods leading to diminished results. This information directly contradicts the

Table 3.5 Top 5 PSTE effect sizes

Year	Author(s)	Innovative practices	Effect size
2003	Eshach	Problem-based learning, curriculum development, inquiry learning, cooperative learning	1.3
2001	Roberts, Henson, Tharp and Moreno	Constructivism, mentoring/in-school support, inquiry learning	1.3
2002	Posnanski	Modelling and vicarious experiences, deep links to mathematics (STEM), field experiences, curriculum development, inquiry learning, school partnership	1.045
2013	Batiza et al.	N/A—Deep content focus	1.03
2001	Kean and Enochs	Field Experience, mentoring, site based learning, tertiary partnership, cooperative learning, ICT instruction, student-centred investigation	0.95

findings of Posnanski (2002). Posnanski described a detailed and scaffolded science program wherein practicing teachers attended 3–4 h weekly meetings over a 32-week period. The participants developed science programs, STEM knowledge, and a repertoire of inquiry pedagogies. The practices were incorporated into a reflective framework where participants' previous practices were critiqued and adjusted, rather than dismantled. This ultimately culminated in a large PSTE effect size growth (Cohen's D = 1.045) for the 31 teachers. The creators of the Students Understanding Energy (SUN) project eschewed pedagogical variance for comprehensive coverage of core science content knowledge (Batiza et al. 2013). Teachers developed their capacity to teach biological energy transfer through hands-on, model building experiences. The participants displayed durable improvements in both their content knowledge scores and personal science teaching efficacy beliefs. Kean and Enochs (2001) found that participation in a 3-week Geology workshop, incorporating student centred investigation and curriculum development opportunities, covaried with large effect size increases in participants' PSTE scores. The next paragraph describes the pedagogical trends amongst the top five STOE interventions.

There were no consistent pedagogical trends amongst the science interventions that showed the greater effects on participants' science teaching outcome expectancies. Table 3.6 ranks the top five interventions in terms of STOE effect sizes produced. The most common innovative practices amongst these interventions were cooperative learning (4) and inquiry learning (4). None of the five interventions incorporated practical teaching experiences. The mean number of innovations (4) was slightly smaller than the mean for all 70 coded interventions (4.6). Perhaps clear, deep science interventions are more appropriate for in-service teachers than complex, pedagogically diverse programs. Deeper research, beyond the constraints of this review, is required to develop a clearer understanding of how the STOE effect sizes of in-service teachers can be improved. While the dichotomous coding and effect size calculation provide a summary of the effectiveness of science

Table 3.6 Top 5 STOE effect sizes

Year	Author(s)	Innovative practices	Effect size
2003	Eshach	Problem-based learning, curriculum development, inquiry learning, cooperative learning	3.1[a]
2013	Ulmer and others	Student-centred investigation	0.806
2011	Holden, Groulx, Bloom and Weinburgh	Constructivism, site based learning, cooperative learning	0.557
2010	Gosselin et al.	Constructivism, site based learning, inquiry learning, tertiary partnership, cooperative learning, ICT instruction, authentic learning, nature of science focus	0.41
2010	Shin et al.	Problem-based learning, mentoring, inquiry learning, cooperative learning, alternative conception targeting	0.383

[a]Denotes outlier

interventions, variables such as the teacher, the pedagogical combinations and the teaching contexts of participants need to be given further consideration.

Amidst the wide-spread struggles to affect change on the STOE subscale, Eshach (2003) reported on a series of inquiry-event based science workshops that produced gargantuan STOE effect size growth (Cohen's D = 3.1) that has never been replicated. The inquiry events required the participants to engage in focused scientific research which was explicitly linked to common community issues. Clearly, developing teachers' understandings of the nature of science and its place in society through successful scientific investigative experiences can improve their science teaching outcome expectancies. Perhaps understanding the nature of science is needed to influence the STOE subscale more so than pedagogical and content development. Ulmer et al. (2013) described an investigative science professional development program where the learning experiences were framed by problem-based learning scenarios. The 88 participating teachers were provided with agricultural science curriculum materials to use within their classrooms. Initial results showed that the participating teachers experienced large STOE growth (Cohen's D = 0.806). However, these gains were not durable as their STOE scores had returned to pre-intervention levels just nine months after the conclusion of the program.

More pedagogically diverse programs have shown to covary with moderate STOE effect size gains for participants. Gosselin et al. (2010) investigated the influence of the Laboratory Earth science professional development program. The complex online program used cooperative learning strategies to guide inquiry learning opportunities. The program afforded participating teachers the opportunity to establish links with the scientific community to help them better understand the nature of both science and investigation. The 51 participants reported improved content knowledge and greater enjoyment of science teaching. Their outcome expectancy beliefs showed statistically significant, albeit small-to-moderate, effect size increases. Shin et al. (2010) reported on a similarly supportive long-term professional development program that utilised reflexive face-to-face workshops rather than online study modules. The innovative practices included: problembased learning, mentoring, inquiry learning, cooperative learning and misconception targeting. The three cohorts display improved science teaching pedagogical repertoires and small-to-moderate STOE increases (Cohen's D = 0.393). The following section will discuss and analyse some of the more prominent findings that emerged from this STEBI-A review.

Discussion

The STEBI-A instrument has been adapted for a variety of compelling research purposes since it was developed and published by the seminal authors (Riggs and Enochs 1990). A simple analysis of the literature in terms of the number of STEBI-A adminstrations with each research project shows a relatively even spread of research approaches, with 40 % of research reporting one use, 23 % reporting two uses and a further 37 % reporting more than two uses of the instrument. The instrument has been used to: conduct deep case study research (Nafziger 2008); target research participants (Ramey-Gassert et al. 1996); make comparisons with student variables (Saam et al. 1999); explore relationships between STEBs and science content knowledge (Lekhu 2013); assess the influence of professional development programs (Nelson 2006); and determine the durability of efficacious change (Sandholtz and Ringstaff 2011). The STEBI-A instrument has had, and continues to have, a lasting impact as a means of consistently measuring efficacy constructs in a field with inherently intangible variables. The STEBI-A instrument has also played a role in the globalisation of science teacher research as nations such as China, Ecuador, The Netherlands and The United Arab Emirates have all contributed to the body of literature within the past five years. Researchers are making strides to capture strong samples of in-service teachers as there was an average of 80 participants per research project. The average only dropped to 65 when the cross sectional and qualitative research was removed. There are still some issues with sampling for research designs with multiple STEBI-A administrations.

Some contextual and methodological issues have emerged from within the STEBI-A literature base. Factors such as: small sample sizes (e.g. Britton 2013), subscale merging (e.g. Rudman and Webb 2009), STOE omission (e.g. Clark 2009) and unclear descriptive statistics serve to mitigate the comparisons of research across different contexts. A number of strong research projects were eliminated at the effect size stage of the analysis because the low number of participants meant that key statistical assumptions could not be met (e.g. McConnell et al. 2008; Mentzer et al. 2014). The limited generalisability of low participant numbers is also an issue for the research that aims to bridge gaps between in-service and pre-service teaching. There are few examples of research that follow the science teaching efficacy beliefs trends in a statistically generalisable fashion. The STEBI-A literature needs to continue to shift from an exploratory focus to an evaluative focus. That is to say that, due to the declining state of science education globally, research needs to consider how science teaching efficacy deficits can be addressed. The use of evaluative research designs with multiple STEBI-A administrations, is a viable and comparable method of assisting the literature in developing along such a pathway. Admittedly, problems with consistent access to teacher data and issues with resentful demoralisation occurring in true experimental designs could hinder such progress. Resentful demoralisation could potentially be addressed by offering modified interventions, rather than no interventions, for control groups (e.g. Fishman et al. 2013).

The PSTE and STOE subscales are generally applied consistently throughout the STEBI-A literature. The PSTE subscale is commonly favoured over the STOE subscale due to its higher reported reliability (Yang et al. 2014) and greater capacity for growth (e.g. Haney et al. 2007). The STOE subscale has similar issues in the STEBI-A literature as are shown in the STEBI-B literature. The STOE subscale is commonly merged inappropriately with PSTE items (e.g. Shea et al. 2013) or omitted entirely from research (e.g. Lucero et al. 2013). When the STOE subscale is measured accurately, it commonly produces lower mean scores (e.g. Liu et al. 2008) and shows less growth (e.g. Ulmer et al. 2013) than the PSTE subscale. In fact, there was a notable trend of small declines in teachers' STOE scores within the STEBI-A literature (Haney et al. 2007; Lockman 2006; Saka et al. 2009). Compelling arguments against the STOE subscale have been made, based on issues such as low reliability (e.g. Nelson 2006), the external locus of construct influence (e.g. Lardy 2011) and even the lack of consistency in how the subscale should be interpreted (Roberts et al. 2001; Settlage et al. 2009). Such arguments have undeniable merit but the outcome expectancies of teachers still need to be considered and addressed. In a practical sense, an individual teacher cannot be extrapolated from the broad social, cultural and socio-economic contexts within which they practice. Without a sufficient belief that potentially detrimental factors can be overcome by science teaching in general, a teacher is less likely to show resilience in challenging professional situations. Therefore, it is inappropriate and short-sighted to place greater emphasis on teachers' personal science teaching efficacy over their more general outcome expectancies.

The conceptual separation of the PSTE and STOE subscales is a key point of differentiation between the STEBI-B and STEBI-A literature. It appears as though the correlation between the PSTE and STOE subscales declines with teaching experience. For the STEBI-B instrument there was a moderate-to-strong correlation between the PSTE and STOE effect sizes produced within research projects (Pearson's $R = 0.628$). However, there was no correlation between the PSTE and STOE effect sizes produced in the STEBI-A research (Pearson's $R = 0.121$). An interpretation may be that entering the teaching profession serves as a catalyst for a conceptual separation of the PSTE and STOE subscales. This is not the first separation of the STEBI-A subscales that has been noted in the literature. Desouza et al. (2004) found that the correlation between the subscales weakened as the level of teacher education attainment increased. One may consider this a process of natural maturation as individuals move from idealistic to realistic conceptualisations of science teaching. If anything, this subscale separation affirms the importance of tertiary science education in establishing core science teaching efficacy beliefs. It also hints at the differing needs of pre-service and in-service teachers. Curiously, there is some evidence that the STOE subscale could be more malleable for in-service teachers as a greater proportion of the STEBI-A literature reports higher growth than the STEBI-A literature. Raising outcome expectancies, adapting to and considering the professional contexts of teachers and explicit consideration of extraneous teaching variables could be used within science teaching professional development programs.

The practical needs of teachers are evident in the pedagogical inclusions within the analysed science programs. Inquiry learning, cooperative group work and curriculum development were the most common pedagogical inclusions amongst the 76 coded science programs. Each of the other innovations was represented within the literature. The same teacher-centred, practical approaches were evident in the interventions that covaried with the strongest PSTE and STOE effect size gains. The mean effect size gain for the PSTE subscale (Cohen's D = 0.572) was notably, if expectedly, larger than the mean effect size gain on the STOE subscale (Cohen's D = 0.398). Curiously, despite the aforementioned separation of the subscales, there were no key pedagogical differences between the top interventions in terms of PSTE and STOE growth. The absence of significant STOE gains throughout the STEBI-A literature more broadly, dilutes the comparison between the subscales in terms of pedagogical influence. There was a slight trend toward less complex program designs with fewer innovative practices. Perhaps clearer program designs account for the need of practicing teachers to balance their professional practice with participation in science professional development. Concerted efforts are being made to overcome issues stemming from difficulty in accessing and supporting in-service teachers. Since 2010 eight programs have incorporated online elements as a means of supplementing traditional face-to-face deliveries (e.g. Sang et al. 2012).

There are a number of implications that have arisen from the findings of this comprehensive STEBI-A review. Firstly, the STEBI-A literature needs to continue to shift from an exploratory focus to an evaluative focus due to the waning state of science education globally. More research with multiple administrations of the STEBI-A instrument and sufficient participant samples is needed to assess the influence of different pedagogical approaches on in-service teachers' science teaching efficacy beliefs. This is particularly important given the differences between the needs of pre-service and in-service teachers. Secondly, the conceptual separation of personal science teaching efficacy and the more general outcome expectancies that appears to be evident for in-service teachers indicates that different approaches to science professional development are needed. When considered in conjunction with the noted stagnation of the STOE scores, it appears that the specific teaching contexts and beliefs of practicing teachers need to be overtly considered in more reflexive approaches to science professional development. Thirdly, the complementary nature of the STEBI-B and STEBI-A instruments affords a unique opportunity for statistically valid and reliable tracking of prospective teachers' science teaching efficacy beliefs as they transition into their professional teaching careers. There needs to be more research conducted to determine if gains to teachers' science teaching efficacy made at the tertiary level remain durable after graduation. Currently, this transition has only been covered in a qualitative manner (e.g. Saka 2007). Finally, online delivery and support for science professional development programs need to be addressed within the STEBI-A literature. Distance support is a viable means for remedying the issues of access and time restrictions that are specific to in-service teachers. In conclusion, the STEBI-A continues to be a valid and reliable instrument that can provide contextually transferable data and insights in an area of science education where effects and changes are difficult to identify and measure.

References

Albion, P. R., & Spence, K. G. (2013). Primary connections in a provincial queens-and school system: relationships to science teaching self-efficacy and practices. *International Journal of Environmental and Science Education, 8*(3), 501–520.

Andersen, A. M., Dragsted, S., Evans, R. H., & Sørensen, H. (2004). The relationship between changes in teachers' self-efficacy beliefs and the science teaching environment of danish first-year elementary teachers. *Journal of Science Teacher Education, 15*(1), 25–38.

Angle, J., & Moseley, C. (2009). Science teacher efficacy and outcome expectancy as predictors of students' End-of-Instruction (EOI) Biology I test scores. *School Science and Mathematics, 109*(8), 473–483.

Azer, S. A. (2001). Problem-based learning. *Saudi Medical Journal, 22*(4), 299–305.

Batiza, A. F., Gruhl, M., Zhang, B., Harrington, T., Roberts, M., LaFlamme, D., & Nelson, D. (2013). The effects of the SUN project on teacher knowledge and self-efficacy regarding energy transfer are significant and long-lasting: Results of a randomised controlled trial. *CBE-Life Sciences Education, 12*(2), 287–305.

Britton, P. S. (2013). *Middle level science: A mixed-methodology study of the impact of the Pennsylvania system of school assessment (PSSA) on curriculum and instruction* (Doctoral dissertation, Indiana University of Pennsylvania).

Burton, L. D. (1996). *How teachers teach: Seventh- and eighth-grade science instruction in the USA.* Paper presented at the Annual Conference of the National Science Teachers Association, St. Louis, MO.

Buyuktaskapu, S. (2010). Examination of pre-school teachers' beliefs about science education. *The International Journal of Research in Teacher Education, 1*, 14–25.

Clark, I. (2009). *An analysis of the relationship between K-5 elementary school teachers' perceptions of principal instructional leadership and their science teaching efficacy* (Doctoral dissertation, University of Minnesota).

Desouza, J. M. S., Boone, W. J., & Yilmaz, O. (2004). A study of science teaching self-efficacy and outcome expectancy beliefs of teachers in India. *Science Education, 88*(6), 837–854.

Duran, E., Ballone-Duran, L., Haney, J., & Beltyukova, S. (2009). The impact of a professional development program integrating informal science education on early childhood teachers' self-efficacy and beliefs about inquiry-based science teaching. *Journal of Elementary Science Education, 21*(4), 53–70.

Duran, E., Ballone-Duran, L., & Haney, J. (2010). Project ASTER III: A model for teacher professional development integrating science museum exhibits with state and national science education content standards. *Curator: The Museum Journal, 53*(4), 437–449.

Edelson, D. C., Gordin, D. N., & Pea, R. D. (1999). Addressing the challenges of inquiry-based learning through technology and curriculum design. *Journal of the Learning Sciences, 8*(3–4), 391–450.

Ellins, K. K., Snow, E., Olson, H. C., Stocks, E., Willis, M., Olson, J., & Odell, M. R. (2013). The Texas Earth and Space Science (TXESS) revolution: A model for the delivery of Earth Science professional development to minority-serving teachers. *Journal of Geoscience Education, 61*(2), 187–201.

Enochs, L. G., & Riggs, I. M. (1990). Further development of an elementary science teaching efficacy belief instrument: A preservice elementary scale. *School Science and Mathematics, 90*(8), 694–706.

Ertmer, P. A., Schlosser, S., Clase, K., & Adedokun, O. (2014). The grand challenge: Helping Teachers learn/teach cutting-edge science via a PBL approach. *Interdisciplinary Journal of Problem-based Learning, 8*(1), 1–17.

Eshach, H. (2003). Inquiry-events as a tool for changing science teaching efficacy belief of kindergarten and elementary school teachers. *Journal of Science Education and Technology, 12*(4), 495–501.

Fathi-Azar, E. (2002). Elementary teachers' science self efficacy beliefs in the East Azerbaijan province of Iran. *Journal of Science and Mathematics Education in Southeast Asia, 25*(1), 95–106.

Finson, K. D., Beaver, J. B., & Hall, L. (1992). Peer mentorship program in scientific literacy for rural elementary teachers. *Journal of Elementary Science Education, 4*(2), 35–48.

Fishman, B., Konstantopoulos, S., Kubitskey, B. W., Vath, R., Park, G., Johnson, H., & Edelson, D. C. (2013). Comparing the impact of online and face-to-face professional development in the context of curriculum implementation. *Journal of Teacher Education, 64*(5), 426–438.

Ginns, I.S., & Watters, J. (1996) *Experiences of novice teachers: Change in self-efficacy and their beliefs about teaching.* Paper presented at the Annual Meeting of the American Educational Research Association, New York, 8–12 April, ERIC Document Reproduction Service, ED 400 243.

Ginns, I. S., & Watters, J. J. (1999). Beginning elementary school teachers and the effective teaching of science. *Journal of Science Teacher Education, 10*(4), 287–313.

Gosselin, D., Thomas, J., Redmond, A., Larson-Miller, C., Yendra, S., Bonnstetter, R., & Slater, T. (2010). Laboratory earth: A model of online K-12 teacher coursework. *Journal of Geoscience Education, 58*(4), 203–213.

Green, J. L., Camilli, G., & Elmore, P. B. (Eds.). (2006). *Handbook of complementary methods in education research.* Routledge.

Haney, J. J., Wang, J., Keil, C., & Zoffel, J. (2007). Enhancing teachers' beliefs and practices through problem-based learning focused on pertinent issues of environmental health science. *The Journal of Environmental Education, 38*(4), 25–33.

Hodgin, C. M. (2014). *Science teaching anxiety: The impact of beliefs on teacher preferences of instructional strategies* (Doctoral dissertation, University of Texas).

Holbert, K. E., Molyneaux, K., Grable, L. L., & Dixon, P. (2011). Multi-university precollege outreach from a renewable energy focused Engineering Research Center. In *Proceedings of ASEE/IEEE Frontiers in Education Conference*, 1–7 retrieved from http://fie-conference.org/fie2011/papers/1083.pdf

Holden, M. E., Groulx, J., Bloom, M. A., & Weinburgh, M. H. (2011). Assessing teacher self-efficacy through an outdoor professional development experience. *Electronic Journal of Science Education, 15*(2), 1–25.

Kahn, P., & O'Rourke, K. (2005). Understanding enquiry-based learning. In T. Barrett, I. M. Labhrainn & H. Fallon (Eds.), *Handbook of Enquiry- and Problem- based learning: Irish case studies and international perspectives* (pp. 1–12). Dublin: Centre for Excellence in Learning and Teaching, NUI Galway and All Ireland Society for Higher Education (AISHE).

Kean, W. F., & Enochs, L. G. (2001). Urban field geology for K-8 teachers. *Journal of Geoscience Education, 49*(4), 358–363.

Kervin, L., Wilma, V., Herrington, J., & Okely, T. (2006). *Research for educators.* Melbourne: Cengage learning Australia.

Khourey-Bowers, C., & Fenk, C. (2009). Influence of constructivist professional development on chemistry content knowledge and scientific model development. *Journal of Science Teacher Education, 20*(5), 437–457.

Kuchey, D., Morrison, J. Q., & Geer, C. H. (2009). A professional development model for math and science educators in Catholic elementary schools: Challenges and successes. *Catholic Education: A Journal of Inquiry and Practice, 12*(4), 475–497.

Lardy, C. H. (2011). *Personal science teaching efficacy and the beliefs and practices of elementary teachers related to science instruction* (Doctoral Dissertation, University of California at San Diego).

Lawrance, G. A., & Palmer, D. A. (2003). *Clever teachers, clever sciences: Preparing teachers for the challenge of teaching science, mathematics and technology in 21st Century Australia.* Canberra: Australian Government, Department of Education, Science and Training: Research Analysis and Evaluation Group.

Lekhu, M. A. (2013). Relationship between self-efficacy beliefs of science teachers and their confidence in content knowledge. *Journal of Psychology in Africa, 23*(1), 109–112.

Liu, C. J., Jack, B. M., & Chiu, H. L. (2008). Taiwan elementary teachers' views of science teaching self-efficacy and outcome expectations. *International Journal of Science and Mathematics Education, 6*(1), 19–35.

Lockman, A. S. (2006). *Changes in teacher efficacy and beliefs during a one-year teacher preparation program* (Doctoral dissertation, The Ohio State University).

Lucero, M., Valcke, M., & Schellens, T. (2013). Teachers' beliefs and self-reported use of inquiry in science education in public primary schools. *International Journal of Science Education, 35*(8), 1407–1423.

McConnell, T. J., Lundeberg, M. A., Koehler, M. J., Urban-Lurain, M., Zhang, T., Mikeska, J., & Eberhardt, J., et al. (2008). *Video-based teacher reflection—What is the real effect on reflections of inservice teachers?* Paper presented at the annual meeting of the Association of Science Teacher Education, St. Louis, MO.

McKinnon, M., Moussa-Inaty, J., & Barza, L. (2014). Science teaching self-efficacy of culturally foreign teachers: A baseline study in Abu Dhabi. *International Journal of Educational Research, 66*, 78–89.

Mentzer, G. A., Czerniak, C. M., & Struble, J. L. (2014). Utilizing program theory and contribution analysis to evaluate the development of science teacher leaders. *Studies in Educational Evaluation, 42*(1), 100–108.

Mulholland, J., Dorman, J. P., & Odgers, B. M. (2004). Assessment of science teaching efficacy of preservice teachers in an Australian university. *Journal of Science Teacher Education, 15*(4), 313–331.

Nadelson, L. S., Callahan, J., Pyke, P., Hay, A., Dance, M., & Pfiester, J. (2013). Teacher STEM perception and preparation: Inquiry-based STEM professional development for elementary teachers. *The Journal of Educational Research, 106*(2), 157–168.

Nafziger, K. M. (2008). *Particulate nature of matter, self-efficacy, and pedagogical content knowledge: Case studies in inquiry* (Doctoral dissertation, Miami University).

Naizer, G., Bell, G. L., West, K., & Chambers, S. (2003). Inquiry science professional development combined with a science summer camp for immediate application. *Journal of Elementary Science Education, 15*(2), 31–37.

Nelson, D. (2006). *Effect of a Concentrated In-service Elementary Teacher Force and Motion Workshop* (Masters thesis, The University of Maine).

Ogbomo, Q. O. (2010). *Science museums, centers and professional development: Teachers self reflection on improving their practice* (Doctoral dissertation, Indiana State University).

Palmer, D. (2011). Sources of efficacy information in an inservice program for elementary teachers. *Science Education, 95*(4), 577–600.

Pinnell, M., Rowly, J., Preiss, S., Franco, S., Blust, R., & Beach, R. (2013). Bridging the gap between engineering design and PK-12 curriculum development through the use the STEM education quality framework. *Journal of STEM Education: Innovations and Research, 14*(4), 28.

Posnanski, T. J. (2002). Professional development programs for elementary science teachers: An analysis of teacher self-efficacy beliefs and a professional development model. *Journal of Science Teacher Education, 13*(2), 189–220.

Ramey-Gassert, L., Shroyer, M. G., & Staver, J. R. (1996). A qualitative study of factors influencing science teaching self-efficacy of elementary level teachers. *Science Education, 80*(3), 283–315.

Riggs, I. M., & Enochs, L. G. (1990). Toward the development of an elementary teacher's science teaching efficacy belief instrument. *Science Education, 74*(6), 625–637.

Roberts, J. K., Henson, R. K., Tharp, B. Z., & Moreno, N. P. (2001). An examination of change in teacher self-efficacy beliefs in science education based on the duration of inservice activities. *Journal of Science Teacher Education, 12*(3), 199–213.

Rudman, N., & Webb, P. (2009). Self-efficacy and the recognition of prior learning. *Africa Education Review, 6*(2), 283–294.

Saint, H. H. (2013). *Personality and science training as predictors of science teaching efficacy beliefs* (Doctoral dissertation, Liberty University).

Saka, Y. (2007). *Exploring the interaction of personal and contextual factors during the induction period of science teachers and how this interaction shapes their enactment of science reform* (Doctoral dissertation, Florida State University).

Saka, Y., Southerland, S. A., & Golden, B. (2009). *Describing the effects of research experiences for teachers attitudes and beliefs: Examining science teachers' self-efficacy beliefs, beliefs of reformed-science teaching, attitudes and pedagogical discontentment*. Paper presented at the annual meeting of the American Educational Research Association, San Diego, CA.

Saam, J., Boone, W. J., & Chase, V. (1999). *A snapshot of upper elementary and middle School science teachers' self-efficacy and outcome expectancy*. Proceedings of the Annual International Conference of the Association for the Education of Teachers in Science, Austin, TX.

Sandholtz, J. H., & Ringstaff, C. (2011). Reversing the downward spiral of science instruction in K-2 classrooms. *Journal of Science Teacher Education, 22*(6), 513–533.

Sandholtz, J. H., & Ringstaff, C. (2014). Inspiring instructional change in elementary school science: The relationship between enhanced self-efficacy and teacher practices. *Journal of Science Teacher Education, 25*(6), 729–751.

Sang, G., Valcke, M., van Braak, J., Zhu, C., Tondeur, J., & Yu, K. (2012). Challenging science teachers' beliefs and practices through a video-case-based intervention in China's primary schools. *Asia-Pacific Journal of Teacher Education, 40*(4), 363–378.

Schmidt, D. L., Saigo, B. W., & Stepans, J. I. (2006). *Conceptual change model: The CCM handbook*. Saiwood Publications.

Settlage, J., Southerland, S. A., Smith, L. K., & Ceglie, R. (2009). Constructing a doubt-free teaching self: Self-efficacy, teacher identity, and science instruction within diverse settings. *Journal of Research in Science Teaching, 46*, 102–125.

Shea, L. M., Shanahan, T. B., Gomez-Zwiep, S., & Straits, W. (2013). Using science as a context for language learning: Impact and implications from two professional development programs. *Electronic Journal of Science Education, 16*(2), 1–29.

Shin, T. S., Koehler, M. J., Lundeberg, M. A., Zhang, M., Eberhardt, J., Zhang, T., & Paik, S. (2010). *The impact of problem-based learning professional development on science teachers' self-efficacy and their teaching practices*. Paper presented at the Annual meeting of American Educational Research Association, Denver, CO.

Shroyer, M. G., Wright, E. L., & Ramey-Gassert, L. (1996). An innovative model for collaborative reform in elementary school science teaching. *Journal of Science Teacher Education, 7*(3), 151–168.

Slavin, R. (1991). Synthesis of research on cooperative learning. *Educational Leadership, 48*(5), 71–82.

Suter, W. N. (2006). *Introduction to educational research: A Critical thinking approach*. Thousand Oaks: Sage Publications.

Taylor, B. K. (2005). *Analysis of environmental and general science efficacy among instructors with contrasting class ethnicity distributions: A four dimensional assessment* (Doctoral dissertation, Oklahoma State University).

Thomas, J., Ivey, T., & Puckette, J. (2013). Where is earth science? Mining for opportunities in chemistry, physics, and biology. *Journal of Geoscience Education, 61*(1), 113–119.

Ulmer, J. D., Velez, J. J., Lambert, M. D., Thompson, G. W., Burris, S., & Witt, P. A. (2013). Exploring science teaching efficacy of CASE curriculum teachers: A post-then-pre assessment. *What a Degree in Agricultural Leadership Really Means: Exploring Student Conceptualizations, 54*(4), 121–133.

Urquhart, M. L., & Bober, K. M. (2006, February). *The impact of teacher quality grants on long-term professional development of physical science teachers*. Paper present at the 2005 Physics Education Research Conference, 818, 27–30.

Velthuis, C. (2014). *Collaborative curriculum design to increase science teaching self-efficacy* (Doctoral dissertation, University of Twente, Enschede).

Vygotsky, L. (1977). The development of higher psychological functions. *Soviet Psychology, 16*(1), 60–73.

Watters, J. J., & Ginns, I. S. (1997). An in-depth study of a teacher engaged in an innovative primary science trial professional development project. *Research in Science Education, 27*(1), 51–69.

Wingfield, M. E., Freeman, L., & Ramsey, J. (2000). *Science teaching self-efficacy of first year elementary teachers trained in a site based program.* Paper presented at the Annual Meeting for the National Association for Research in Science Teaching, New Orleans, LA.

Yang, C., Noh, T., Scharmann, L. C., & Kang, S. (2014). A study on the elementary school teachers' awareness of students' alternative conceptions about change of states and dissolution. *The Asia-Pacific Education Researcher, 23*(3), 683–698.

Chapter 4
Conclusion

The analytic framework presented in this Springer Brief has allowed for the deep and systematic analysis of over 240 articles and dissertations which have used the Science Teaching Efficacy Belief Instruments (A-B) over the past 25-years. The separate analyses of research methods and findings has allowed for the relatively objective critique and organisation of a large (combined citation score of 1400) and rapidly growing body of literature. While the increasing availability of online publication platforms has enhanced the sum of science education knowledge, it presents new challenges to modern education researchers as they strive to reconcile their own work within increasingly expansive fields of research. Thus, it stands to reason that meta analytic and review research will need to be conducted to ensure the research community continues to respond to appropriate gaps in knowledge based on changing needs, to ultimately affect positive change in science education. The need for the critique and summary of burgeoning literature bodies extends to other fields of science education and, indeed, educational research in general. Aside from providing a thorough critique and organised summary of the STEBI-B/A literature bodies, it is hoped that the analytic framework presented in this Springer Brief can serve as a model for future review research in other domains. At the very least, the reader should spare a thought for the PhD candidates, teachers, policy makers and educational administrators who are likely to be overwhelmed by the increasing permeation of valuable education research.

The STEBI-B instrument has been used in varied ways to advance science education research. The United States of America, Australia and Turkey are the current global leaders in the production of STEBI-B research with increasing contribution rates from other nations. The STEBI-B research methods are similarly diverse. Qualitative STEBI-B research has served to establish the importance of teacher efficacy in science education. Tosun (2000) modified the STEBI-B instrument to create interview questions for 17 pre-service teachers. Participants' responses to their tertiary science experiences were overwhelmingly negative, despite notable differences in achievement scores. This indicates that teachers' feelings towards science may overshadow academic achievement in terms of influence on science teaching efficacy beliefs. A similar qualitative study revealed

© The Author(s) 2017
J. Deehan, *The Science Teaching Efficacy Belief Instruments (STEBI A and B)*,
SpringerBriefs in Education, DOI 10.1007/978-3-319-42465-1_4

that pre-service teachers place a greater emphasis on "fun" rather than academic rigour when selecting and developing science lessons (Peters-Burton and Hiller 2013). Cross sectional research indicates positive correlations amongst pre-service teachers' science teaching efficacy beliefs and other variables, including: classroom management beliefs (Gencer and Çakiroglu 2007), epistemological views (Yilmaz-Tuzun and Topcu 2008) and science content knowledge (Sarikaya et al. 2005). In recent years, more complex research designs, with multiple STEBI-B administrations, have been used to assess the efficacy of different approaches to tertiary science education. Results have indicated approaches such as curriculum development (Jabot 2002), mentoring (Cooper 2015), cooperative learning (Palmer 2006), in-subject practical teaching experience (Bautista 2011) and links to extended professional placement blocks (Swars and Dooley 2010) can all be viable ways of enhancing the science teaching efficacy beliefs of prospective teachers. There is now overwhelming evidence to remove transmissive, disengaging, content-heavy approaches from pre-service science education programs and courses.

The STEBI-A literature exhibits similar variance in terms of methodologies and findings. Most of the research originates from the USA with a small, but meaningful, group of contributions emerging from other nations. Modified, qualitative use of the STEBI-A has deepened our understanding of science teaching efficacy, relevant antecedents and teachers' science teaching experiences. In fact, qualitative STEBI-A research has highlighted the influence of school contexts on teachers' science teaching efficacy beliefs (Ginns and Watters 1996) and the apparent dissonance between tertiary education messages and professional teaching contexts (Saka 2007). Such antecedents will undoubtedly inform new STEBI research. Cross sectional STEBI-A research has broadened the relevance of science teaching efficacy by establishing links to: the use of constructivist teaching practices (Burton 1996), preferences for affording student autonomy (Lucero et al. 2013), and the time spent teaching science (Albion and Spence 2013). A significant section of quasi experimental STEBI-A research has provided empirical evidence to prove the worth of approaches such as inquiry learning (Eshach 2003), mentoring/in-school support (Roberts et al. 2001), curriculum development (Posnanski 2002) and tertiary partnerships (Kean and Enochs 2001) for improving the science teaching efficacy beliefs of in-service teachers. While there is a predictable reliance on face-to-face modes of communication across the analysed professional development programs, many practitioners and researchers have addressed issues of time and geographical disparity through in-school follow up visits and opportunities for general course reflection (Shin et al. 2010). In recent years, online learning has become a more prominent inclusion within science education professional development programs as a means of fostering more direct communication between instructors and in-service teachers (Blackmon 2003; Haeusler and Lozanovski 2010; Sang et al. 2012). Diverse, learner-centred programs as such "Primary Connections" (Hackling 2006) and "Science by Doing" (Rennie 2010) are examples of beneficial changes that are occurring within the field of in-service science teacher education.

STEBI analyses revealed several similarities between pre-service and in-service teachers. In terms of science education program design, there are no clear differences between pre-service and in-service teachers in the innovative practices that have shown to covary with increased science teaching efficacy beliefs. After 25-years of global STEBI research, there is ample evidence for the use complex, cooperative and learner-centred practices for science teacher education (Lumpe et al. 2012; Palmer 2006). Many prospective and practising teachers have reported greater increases to their personal science teaching efficacy belief scores (Bautista and Boone 2015; Nelson 2006), even though the PSTE subscale has shown to produce consistently higher mean scores than the Science Teaching Outcome Expectancy (STOE) subscale (Bautista 2011; Saint 2013) Some disparity between the subscales is to be expected given the general, more open nature of the STOE subscale. However, this does not remove the onus on instructors and researchers to improve the science teaching outcome expectancies of in-service and pre-service teachers alike. At the very least, the reported decreases in teachers' STOEs need to be considered in the development of future science programs and research projects (Haney et al. 2007; Saka et al. 2009).

The key point of difference between pre-service and in-service teachers is the relationship between personal science teaching efficacy (PSTE) beliefs and science teaching outcome expectancies (STOE). Analysis of STEB effect size changes shows that there is a moderate-to-strong correlation (Pearson's $R = 0.628$) between the subscales for pre-service teachers. Curiously, there is no such correlation between the PSTE and STOE growth scores of in-service teachers (Pearson's $R = 0.121$). When considered in conjunction with the noted challenge of improving the STOEs of practicing teachers (Blackmon 2003; Ewing-Taylor 2012) it would seem that the transition from pre-service to in-service status serves as a catalyst for the separation of the subscales. Such a trend seems unavoidable as school contexts introduce powerful antecedent variables (e.g. colleagues, parents, administrators, etc.) that will likely influence the outcome expectancies of in-service teachers. This is not a new finding as prior research has shown the correlation between the STEB subscales declines as levels of education qualifications increase (Desouza et al. 2004). Perhaps this efficacious separation is related to a process of professional maturation wherein teachers move from idealistic notions of science teaching to hold more realistic, grounded perceptions of the value of science teaching. Still, a portion of the STEBI-B research has shown that preparatory science programs can improve the STOEs of pre-service teachers (Cooper 2015; Ozdelik and Bulunuz 2009; Templeton 2007). Tertiary science education programs must play a key role in improving pre-service teachers' science outcome expectancies to ensure their resilience when faced with potentially detrimental extraneous variables in professional school contexts.

There are a number implications for future research which have emerged from the STEBI-B and STEBI-A reviews contained within this Springer Brief. First, despite the widespread permeation of STEBI research over the past quarter of a century, there are relatively few examples of research using both instruments in a complementary fashion to investigate teachers' transitions from pre-service to in-service status (McKinnon and Lamberts 2013). Such transitional phases represent an open and promising area of science education research. Of particular note is the apparent dissonance between the positive growth trends reported in the tertiary science domain (Bautista 2011; Palmer 2006) and the continued negative reports of teacher attitudes (Jarrett 1999), transmissive pedagogies (Jarvis and Pell 2005), student disengagement (Dewitt et al. 2014) and student achievement (Thomson et al. 2008, 2011). Second, more complex STEBI research designs are needed to assess the durability of STEB changes reported during science interventions. The STEBI instruments have already been used to: explore the nature of science teaching efficacy in individual contexts (Mulholland and Wallace 2003), assess the relationship between science teaching efficacy and other relevant variables (Enochs et al. 1995) and evaluate the impact of innovative approaches to science education (Jabot 2002). Thus, the next logical steps would be to assess the durability of science teaching efficacy growth and, ultimately, explore how science teaching efficacy and science education experience influence classroom teaching practice. Third, the established nature of the STEBI literature base should allow for more connections between different sub-domains within the literature itself. For example, the global use of the STEBI instruments affords the opportunity for rich international research collaboration (Çakiroglu et al. 2005; Rogers and Watters 2002). Fourth, STEBI researchers should begin to consider the modes through which science interventions are delivered. Admittedly, face-to-face education was the deserved focus when the STEBI instruments were published in 1990. However, science education programs are using online and blended modes of delivery at higher rates than ever before (Gosselin et al. 2010; Kean and Enochs 2001). Yet the impact of online and blended modes of delivery on participants' STEBs is still largely unknown. Finally, review research needs become more prominent within science education as rates of research production continue to increase.

To some extent the onus now falls on policy makers, administrators and instructors to act upon the information presented by STEBI researchers to affect change within science education. There is now a plethora of evidence indicating that transmissive, content-heavy science courses should no longer serve as the cornerstone of pre-service and in-service science education programs. Student centred practices such as cooperative learning (Palmer 2006), inquiry learning (Swars and Dooley 2010), alternative conception targeting (Bautista 2011) and problem based learning (Eshach 2003) have been shown to improve participants' science teaching efficacy beliefs across varied time periods and national contexts. Innovative practices will be vital to ensuring teachers are able to foster the scientific literacy of their students as they function in an intermediary role of connecting the general populace with the scientific community. There also appears to be opportunities for universities and colleges to play an expanded role in science education at more fundamental levels. Community and school partnerships (Nadelson et al. 2012; Urquhart and Bober 2006) can be viable means of expanding the influence of student-centred university programs and addressing issues that may arise as teachers transition from undergraduate study into the teaching profession. Online modes of communication promise to make meaningful partnerships between schools and universities more accessible than at any point in history.

As a fledgling researcher and science education academic, it has been both a privilege and an enlightening experience to critique, summarise and organise the research of the STEBI scholars that have preceded me over the past 25-year years. Since the initial publication of the STEBI instruments (Enochs and Riggs 1990; Riggs and Enochs 1990) the body of literature has expanded to over 240 articles and dissertations (and still rapidly increasing); with contributions from over 20 nations making this area of science education research a truly global enterprise. I wholeheartedly believe that our predecessors have laid the foundation necessary to address many of the widespread and evolving challenges to science education that hinder the development of global scientific literacy levels. Education is the omnipresent, yet seldom considered, factor that underpins many of the political, environmental, economic and social challenges at the forefront of our attention in modern times. Therefore, the value of the contributions of teachers, academics, policy makers, educational designers, politicians and other interested parties to science education cannot be overstated. It is my sincerest hope that you, the reader, can extract some value from this Springer Brief. Whether you enter a new branch of science education research, review another body of science education literature, implement broader reforms to your science education practices or, simply, make a minor tweak to the science education in your context, this STEBI review will have achieved the desired purpose. I will be satisfied even if just one doctoral student manages to save some time and/or potential insomnia. I look forward to witnessing the progress of the STEBI literature and making some contributions of my own over the next 25 years.

References

Albion, P. R., & Spence, K. G. (2013). Primary Connections in a provincial Queensland school system: Relationships to science teaching self-efficacy and practices. *International Journal of Environmental and Science Education, 8*(3), 501–520.

Bautista, N. U. (2011). Investigating the use of vicarious and mastery experiences in influencing early childhood education majors' self-efficacy beliefs. *Journal of Science Teacher Education, 22*(4), 333–349.

Bautista, N. U., & Boone, W. J. (2015). Exploring the impact of TeachME™ lab virtual classroom teaching simulation on early childhood education majors' self-efficacy beliefs. *Journal of Science Teacher Education, 26*(3), 237–262.

Blackmon, S. A. (2003). *Empowering elementary teachers in Texas to prepare their students for the science section of the Texas assessment of knowledge and skills (TAKS) 2003* (Doctoral dissertation, Texas A&M University).

Burton, L. D. (1996). *How teachers teach: Seventh- and eighth-grade science instruction in the USA*. Paper presented at the Annual Conference of the National Science Teachers Association, St. Louis, MO.

Çakiroglu, J., Çakiroglu, E., & Boone, W. (2005). Pre-service teacher self-efficacy beliefs regarding science teaching: A comparison of pre-service teachers in Turkey and the USA. *Science Educator, 1*(14), 31–41.

Cooper, T. O. (2015). *Investigating the effects of cognitive apprenticeship-based instructional coaching on science teaching efficacy beliefs* (Doctoral dissertation, Florida International University).

Desouza, J. M. S., Boone, W. J., & Yilmaz, O. (2004). A study of science teaching self-efficacy and outcome expectancy beliefs of teachers in India. *Science Education, 88*(6), 837–854.

DeWitt, J., Archer, L., & Osborne, J. (2014). Science-related aspirations across the primary–secondary divide: Evidence from two surveys in England. *International Journal of Science Education*, (ahead-of-print), 1–21.

Enochs, L. G., & Riggs, I. M. (1990). Further development of an elementary science teaching efficacy belief instrument: A preservice elementary scale. *School Science and Mathematics, 90*(8), 694–706.

Enochs, L. G., Scharmann, L. C., & Riggs, I. M. (1995). The relationship of pupil control to pre-service elementary science teacher self–efficacy and outcome expectancy. *Science Education, 79*(1), 63–75.

Eshach, H. (2003). Inquiry-events as a tool for changing science teaching efficacy belief of kindergarten and elementary school teachers. *Journal of Science Education and Technology, 12*(4), 495–501.

Ewing-Taylor, J. M. (2012). *A longitudinal evaluation study of a science professional development program for K-12 teachers: NERDS* (Doctoral dissertation, University of Nevada, Reno).

Gencer, A. S., & Çakiroglu, J. (2007). Turkish preservice science teachers' efficacy beliefs regarding science teaching and their beliefs about classroom management. *Teaching and Teacher Education, 23*(5), 664–675.

Ginns, I. S., & Watters, J. (1996) *Experiences of novice teachers: Change in self-efficacy and their beliefs about teaching.* Paper presented at the Annual Meeting of the American Educational Research Association, New York, 8–12 April, ERIC Document Reproduction Service, ED 400 243.

Gosselin, D., Thomas, J., Redmond, A., Larson-Miller, C., Yendra, S., Bonnstetter, R., & Slater, T. (2010). Laboratory earth: A model of online K-12 teacher coursework. *Journal of Geoscience Education, 58*(4), 203–213.

Hackling, M. (2006). *Primary Connections: A new approach to primary science and to teacher professional learning.* Paper presented at the ACER Research Conference. Retrieved 13th January 2014 from http://research.acer.edu.au/research_conference_2006/14/.

Haeusler, C. E., & Lozanovski, C. (2010). Enhancing pre-service primary teachers' learning in science education using team-based project work. In *Proceedings of the International Conference on Enhancing Learning Experiences in Higher Education (CETL 2010)* (pp. 1–16). University of Hong Kong.

Haney, J. J., Wang, J., Keil, C., & Zoffel, J. (2007). Enhancing teachers' beliefs and practices through problem-based learning focused on pertinent issues of environmental health science. *The Journal of Environmental Education, 38*(4), 25–33.

Jabot, M. (2002). The effectiveness of a misconceptions-based approach to the teaching of elementary science methods. In *Proceedings of the Pathways to Change International Conference on Transforming Math and Science Education in the K16 Continuum.* Arlington VA: The Science, Technology, Engineering and Mathematics Teacher Education Collaborative (STEMTEC).

Jarrett, O. (1999). Science interest and confidence among preservice elementary teachers. *Journal of Elementary Science Education, 11*(1), 47–57.

Jarvis, T., & Pell, A. (2005). Factors influencing elementary school children's attitudes toward science before, during, and after a visit to the UK National Space Centre. *Journal of Research in Science Teaching, 42*(1), 53–83.

Kean, W. F., & Enochs, L. G. (2001). Urban field geology for K-8 teachers. *Journal of Geoscience Education, 49*(4), 358–363.

Lucero, M., Valcke, M., & Schellens, T. (2013). Teachers' beliefs and self-reported use of inquiry in science education in public primary schools. *International Journal of Science Education, 35*(8), 1407–1423.

Lumpe, A., Czerniak, C., Haney, J., & Beltyukova, S. (2012). Beliefs about teaching science: The relationship between elementary teachers' participation in professional development and student achievement. *International Journal of Science Education, 34*(2), 153–166.

McKinnon, M., & Lamberts, R. (2013). Influencing science teaching self-efficacy beliefs of primary school teachers: A longitudinal case study. *International Journal of Science Education, Part B*, (ahead-of-print), 1–23.

Mulholland, J., & Wallace, J. (2003). Crossing borders: Learning and teaching primary science in the pre-service and in-service transition. *International Journal of Science Education, 25*(7), 879–898.

Nadelson, L. S., Seifert, A., Moll, A. J., & Coats, B. (2012). i-STEM summer institute: An integrated approach to teacher professional development in STEM. *Journal of STEM Education: Innovations and Research, 13*(2), 69.

Nelson, D. (2006). *Effect of a concentrated in-service elementary teacher force and motion workshop* (Masters thesis). The University of Maine.

Ozdelik, Z., & Bulunuz, N. (2009). The effect of a guided inquiry method on pre-service teachers' science teaching self-efficacy beliefs. *Turkish Science Education Journal, 6*(2), 24–42.

Palmer, D. (2006). Durability of changes in self-efficacy of preservice primary teachers. *International Journal of Science Education, 28*(6), 655–671.

Peters-Burton, E. E., & Hiller, S. E. (2013). Fun Science: The Use of Variable Manipulation to Avoid Content Instruction. *Journal of Science Teacher Education, 24*(1), 199–217.

Posnanski, T. J. (2002). Professional development programs for elementary science teachers: An analysis of teacher self-efficacy beliefs and a professional development model. *Journal of Science Teacher Education, 13*(2), 189–220.

Rennie, L. J. (2010). *Evaluation of the science by doing stage one professional learning approach 2010*. Canberra: Australian Academy of Science.

Riggs, I. M., & Enochs, L. (1990). Toward the development of an elementary teachers' science teaching efficacy belief instrument. *Science Education, 74*, 625–637.

Roberts, J. K., Henson, R. K., Tharp, B. Z., & Moreno, N. P. (2001). An examination of change in teacher self-efficacy beliefs in science education based on the duration of inservice activities. *Journal of Science Teacher Education, 12*(3), 199–213.

Rogers, G. & Watters, J.J. (2002). Global perspectives of science education: Successes and challenges of a pilot project. *Paper presented at the Annual Meeting of the Australasian Science Education Research Association Conference*, Melbourne, July 9–12, 2002.

Saint, H. H. (2013). *Personality and science training as predictors of science teaching efficacy beliefs* (Doctoral dissertation, Liberty University).

Saka, Y. (2007). *Exploring the interaction of personal and contextual factors during the induction period of science teachers and how this interaction shapes their enactment of science reform* (Doctoral dissertation, Florida State University).

Saka, Y., Southerland, S. A., & Golden, B. (2009). *Describing the effects of research experiences for teachers attitudes and beliefs: Examining science teachers' self-efficacy beliefs, beliefs of reformed-science teaching, attitudes and pedagogical discontentment*. Paper presented at the annual meeting of the American Educational Research Association, San Diego, CA.

Sang, G., Valcke, M., van Braak, J., Zhu, C., Tondeur, J., & Yu, K. (2012). Challenging science teachers' beliefs and practices through a video-case-based intervention in China's primary schools. *Asia-Pacific Journal of Teacher Education, 40*(4), 363–378.

Sarikaya, H., Çakiroglu, J., & Tekkaya, C. (2005). Self-efficacy, attitude and science knowledge. *Academic Exchange Quarterly, 9*(4), 38–42.

Shin, T. S., Koehler, M. J., Lundeberg, M. A., Zhang, M., Eberhardt, J., Zhang, T., & Paik, S. (2010). *The impact of problem-based learning professional development on science teachers' self-efficacy and their teaching practices*. Paper presented at the Annual meeting of American Educational Research Association, Denver, CO.

Swars, S. L., & Dooley, C. M. (2010). Changes in teaching efficacy during a professional development school-based science methods course. *School Science and Mathematics, 110*(4), 193–202.

Templeton, C. K. (2007). *The impact of a museum-based science methods course on early childhood/elementary pre-service teachers' self-efficacy and ability to develop curriculum using a constructivist approach* (Doctoral dissertation, Florida Atlantic University).

Thomson, S., Wernet, N., Underwood, C., & Nicholas, M. (2008). *TIMSS 2007: Taking a closer look at mathematics and science in Australia.* Melbourne: Australian Council for Educational Research.

Thomson, S., Hillman, K., Wernet, N., Schmid, M., Buckley, S., & Munene, A. (2011). *Monitoring Australian year 4 student achievement internationally: TIMSS and PIRLS 2011.* Melbourne: Australian Council for Educational Research.

Tosun, T. (2000). The beliefs of preservice elementary teachers toward science and science teaching. *School Science and Mathematics, 100*(7), 374–379.

Urquhart, M. L., & Bober, K. M. (2006, February). *The impact of teacher quality grants on long-term professional development of physical science teachers.* Paper present at the 2005 Physics Education Research Conference, 818, 27–30.

Yilmaz-Tuzun, O., & Topcu, M. S. (2008). Relationships among preservice science teachers' epistemological beliefs, epistemological world views, and self-efficacy beliefs. *International Journal of Science Education, 30*(1), 65–85.